Fire!

SURVIVAL &
PREVENTION

Fire!

WILLIAM BLAIR
Fire Commissioner of the City of Chicago

Survival & Prevention

Illustrations by ALLAN PHILLIPS

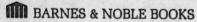 BARNES & NOBLE BOOKS
A DIVISION OF HARPER & ROW, PUBLISHERS
New York, Cambridge, Philadelphia, San Francisco,
London, Mexico City, São Paulo, Sydney

CONTENTS

INTRODUCTION

I'm glad you picked up this book—it means you're concerned about fire. But I want to tell you straight out that unless you're willing to go all the way, unless you're willing to do everything in your power to keep from dying in a fire, I won't have accomplished my purpose in writing it.

I know that people have good intentions. But I also know that people tend to be lazy, that they are apt to think they lead charmed lives. If you are not willing to admit that a life-threatening fire could strike where you are this very day, let me ask you some questions.

Are you sure there are no fire hazards in your house? Are you certain there are enough smoke detectors in your house, and that they all work? Are you positive your family knows exactly what to do in case a fire breaks out despite all your precautions?

If you feel secure simply because there's a fire hydrant in front of your house, or because you've been told your house is fireproof, or because you're young, strong, and athletic—then you have no reason to feel secure. You're being complacent and foolhardy.

Is it really worth it to risk everything you have—including your very life—because you "don't have the time" to sit down with your family for an hour or so and plan what to do should fire break out? Can you honestly say that you can't afford the few dollars that a smoke detector costs? Are you counting on dumb luck to keep you safe from fires?

The truth is, fire does not discriminate. It kills the young as often as the elderly. Strong young men and women in top physical condition can be overcome in a matter of seconds. Children who only a couple of hours earlier were running and playing can die while trying to find a way out of their own bedrooms. Folks who have lived in the same house for forty years can be trapped beyond rescue because their customary exit is blocked by flames.

Many people die in their beds without ever even knowing there is a fire in their homes. And it doesn't have to be a big fire, either. A smoldering cushion in an easy chair is enough to kill—silently, odorlessly, efficiently—before it quietly burns itself out.

Now I suppose this all sounds pretty scary. And I'm not even telling you horror stories about huge disasters in which scores of people died. Such tragedies receive national attention (we all remember the eighty-four lives lost in the Las Vegas MGM Grand Hotel fire a while back). The really heartbreaking fires, though, are those hundreds of small ones that year after year take the lives of innocent people—children who were never shown how to escape from a smoke-filled house; invalids whose bedrooms were heated by space heaters plugged into overloaded outlets; babies whose cribs were in rooms "protected" by burglar bars at the windows; people whose next-door neighbor decided to rinse out some grease-soaked work clothes with a little gasoline.

Of course I'm trying to scare you. If I can get across to you some of the anger and frustration I feel when I see the destruction and grief caused by carelessness; if I can convince you to take some easy, basic steps to minimize the dangers in your home; if I can help you save your life and the lives of your loved ones, then I will have accomplished what I set out to do. Keep in mind that I'm trying to get your attention. I'm trying to show you the dangers connected with fire—some you probably have never thought of—and how to eliminate them. I'm trying to help you find ways to help yourself.

Obviously, I can't come to your home and point out all the hazards around you. I can't devise an escape plan to fit your specific home or tell you exactly how to act in any situation whenever or wherever a fire might start. But I can show you how, with just a little time and effort, you can greatly increase your chances of survival and decrease the chances of a life-threatening fire starting in the first place.

In short, I hope to help you save yourself.

Fire!

SURVIVAL &
PREVENTION

1. GETTING OUT ALIVE

I'm sure that you've heard and read about residential fires all your life and about how individuals and entire families have perished in them. You may have said to yourself, "Why didn't they just get out and let the house burn? How could they have been so foolish? It was just a little grease fire. They could have made it out in no time." It sounds so easy, doesn't it?

We all know how to walk out of a house, and we all know that it's not smart to stick around in a burning building. Why, then, do so many people die in residential fires every year? Why don't they simply leave when their homes start to burn?

The fact is, they don't know the right way to leave. When faced with danger, their first reaction is likely to be to run. But what happens when they run? They make mistakes in judgment, they stumble, they become short of breath and start to gasp, and their excitement makes it hard to think clearly. They forget the rules of fire survival, if they ever knew them. If the fire seems small, they may try to put it out themselves, or take the time to get dressed and gather up some of their possessions. These may be fatal mistakes.

If you're going to live through a fire in your home, you may have to change your way of thinking and reacting to danger. Here are some survival tactics.

Getting Away from a Fire

Feel door for heat.
Brace yourself and open door slowly.
Check hallway for smoke.
If you see smoke, close door and use secondary exit
  (door or window).

If you have no other escape route, crawl under
smoke to exit.

Use stairs only; avoid elevators.

If fire starts in room you are in, leave at once and
close door.

Notify fire department.

Once safely outside, do not reenter for any reason.

Before opening any door, you must be certain there is no
fire on the other side. Opening a door into a fire area is the
worst thing you can do—the fire is likely to rush into the
room before you can get the door closed again. Actually,
flames and smoke won't rush in so much as they will explode
into your room. There is tremendous force behind a fire, and
it can invade a new source of oxygen and fuel faster than you
can slam a door.

To test a door for heat on the other side, feel toward the
top of the door with the back of your hand. Heat rises, and
the top of the door will warm up first. If the door feels cool
but you see smoke curling in at the bottom, don't even consider opening it. That means the area on the other side is
filled to the floor with smoke, and you must find another way
out. Smoke rises and fills a room or a hallway from the top
down.

The right way to open a door is to brace yourself first, one
shoulder against the door and one foot planted firmly about
four or five inches from it. Brace your other foot solidly back
so that you form a wedge against the door. You're not ready
to leave yet, only to look out to see if it's clear and safe. Be
ready to slam the door shut with all your strength—and it may
take all your strength to do it!

Open the door only a few inches—very slowly—and look
out into the hallway. If you see heavy smoke or flames, close
the door immediately. Above all, don't rush out into the

1-1 The right way to open a door.

smoke-filled hall and try to make it to an exit or try to run to another room, either for safety or to warn other members of your family. By this time you'll probably be yelling out warnings anyway. That's good. You should yell, scream, call out names, yell "Fire!" But unless there is no smoke to be seen, don't attempt to help anyone but the helpless and yourself. Your first priority must always be to get out and away from the fire as quickly as possible.

If the hall is filled with smoke, don't go out into it. Instead, slam the door shut and get out through another exit. That will most likely be a window. A great many fires occur at night, so it's very possible you will be in your bedroom when fire hits your home.

A window is not a bad exit if you live in a one- or two-story house; even a third-floor apartment window can be counted as an exit. Don't try it from higher than a third floor. You're better off staying in the room and sealing it from the fire. If you leave through a window, what you must remember is to drop, not jump, to the ground so that your legs can absorb the fall.

When you open the door to the hall, you might see a little smoke drifting up toward the ceiling. If that's the case, your best course of action is to drop to the floor and crawl under the smoke to the nearest exit. Since smoke and gases rise, the clearest and safest air is at floor level. Don't try to go through heavy smoke unless there is absolutely no other way out and your room is also filled with smoke or on fire. Even then, though, you can increase your chances of survival if you get down as low as possible, cover your mouth and nose with a cloth (damp, if possible), and make your way quickly to the exit.

What is an exit? It's a door leading outside or a stairway leading to an outside door. It is NEVER, NEVER an elevator! An elevator, instead of taking you to the ground level, may deliver you straight to the fire floor. This is because in

1-2 Crawl down hall under smoke to nearest exit.

many elevators the passenger signal buttons on each floor are heat-sensitive. The heat from a fire on the eighth floor, for example, will activate the call button and make the car stop at the eighth floor. Once the car stops there and the door opens, you will have absolutely no escape. Also, the operating mechanism could fail during a fire. Whatever the case, the last place you want to be is trapped inside a stalled elevator car in a shaft that is rapidly filling with choking smoke.

If a fire starts in the room you are in, get out fast. It may be a cooking fire, a blazing wastebasket, or flaming draperies ignited by an electrical spark. Whatever it is, don't try to be a hero—get out! As you leave, yell to warn anyone else in the house. Try to confine the fire by closing as many doors as possible as you run out. By doing nothing more than closing a door on the fire room, you may contain the fire in a fairly small area and prevent additional damage or injury to others.

Why haven't I told you yet to call the fire department? Because the most important thing for you to remember and to do is to get out and away from any fire. Your safety and your life are more important than anything else. Once you have made it safely outside, yell to a neighbor or get to a telephone or fire call box and report the exact location of the fire. Taking time to call before trying to escape may mean the difference between surviving and dying. Again, heroics are not called for. What's needed are calm, quick thinking and a fast escape.

Once you're outside and safe, don't *under any circumstances* reenter the burning building. Not even if you think someone may still be inside. Not even if you see only smoke and no flames. Not even if your most precious belongings are still in the house. Not for any reason!

In the few seconds or minutes it took you to reach safety, smoke and poisonous gases have been building up inside. Going back in and taking just one breath could be a fatal mistake. Far too many have died trying to save loved ones who had already escaped but were not seen outside by the would-be rescuer.

The few minutes it takes for a fire-fighting company to

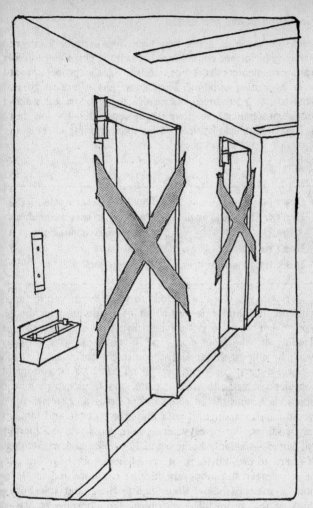

-3 NEVER use an elevator during a fire!

reach the scene of a fire may seem like eternity to anyone watching his or her home burn, but still the best thing to do is wait. Fire fighters have been trained in the proper way to enter a burning building. They know just where to go and what to do. Your loved ones are actually better off waiting for a professional to save them—especially if you have trained them ahead of time. More about training later.

Don'ts

Don't panic.	Don't try to save pets.
Don't try to go through fire.	Don't try to save valuables.
Don't run.	Don't worry about how you look.
Don't try to save others.	Don't go back in.

Very simply, you can't afford to panic when you're faced with the life-threatening situation of a fire in your home. You've got work to do—staying alive isn't a matter of just strolling out the front door—and if you panic, you not only won't be able to think clearly, you might not make it at all.

Sure, it's easy for me to say "Don't panic." And you might consider yourself the kind of calm, cool-headed person who functions beautifully in moments of crisis and stress. What would happen, though, if your most comfortable and familiar surroundings, the place where you're most relaxed—literally "at home"—suddenly became totally foreign and frightening? It's hard to do, but think of your home, your bedroom perhaps, where you know every stick of furniture and the location of every window, door, light switch, and telephone—think of that room filled with black, lung-burning smoke so thick that even if you *could* find the light switch, it wouldn't help you see one bit.

Hard to imagine, isn't it? It's not at all like the game you used to play as a kid when someone would blindfold you and spin you around three times until you didn't know which direction you were going in. It's not a game, and it's scary. Nothing looks the same, smells the same, feels the same. Can you keep from panicking at a time like that?

There is no guarantee, but there is a way—really the only way—that you can prevent that kind of panic. That is by anticipating danger and training yourself so thoroughly ahead of time that your actions, when they are needed, will be automatic. Knowledge and lots of practice can save you and everyone else in your family. So when I say "Don't panic," I'm not being flip or unrealistic. What I'm telling you is that you can increase your chances of survival immeasurably by learning what to do in a fire emergency and learning so well that you won't have time to panic. You'll be too busy saving yourself.

One fundamental point to remember is never to attempt to go through a fire. You might open a door and see a fire burning in the middle of the room. You know the outside door is just beyond the fire—maybe you can even see it through the flames. What you don't know, however, is exactly how much superheated air and poisonous fumes have already collected in the room. You don't know how near to flashover temperature that room is. (Flashover temperature is that temperature at which every bit of combustible material in the room—furniture, draperies, linens, carpets—suddenly explodes into flame.) Also, unknown to you, the outside door may be stuck, or someone may have bolted it or piled something against the other side of it.

Don't take the chance. Find another way out. And do it fast!

Your first inclination in a fire is to run. *Don't.* Instead, walk fast and hold on to stair railings. Even better, crawl. If you're crawling, you not only won't trip and fall, but you'll be breathing cleaner air while you make your way outside.

Unless your household includes infants or bedridden invalids, don't try to save anyone else but yourself. Heartless? Not at all. If you are properly prepared for a fire in your home and have taken the time and responsibility to train everyone else, they will know what to do as well as you do.

Even very small children can be educated about fire dangers and trained to react quickly and effectively to save themselves.

Maybe you have your invalid mother living with you, and maybe she weighs 175 pounds and is helpless. Do you honestly believe you could get her out of bed, through a smoke-filled hallway, down a flight of steps, and out the front door? Very likely, neither one of you would make it, and there would be two needless deaths. The best thing to do in a case like this is to plan and take precautions before a fire hits. A self-closing door with tight weatherstripping on the invalid's bedroom is a good first step. A ceiling-level ventilator or window to allow smoke and gases to escape is another measure you might consider. Unless a fire actually started in her room (and that's the room where you should be most certain that every fire-prevention precaution has been taken), your invalid mother would have a very good chance of surviving until professional fire fighters could rescue her.

People die every year trying to save the lives of their dogs and cats. Maybe I sound like an animal hater (I definitely am not), but I've never met a dog that was worth a human life. The truth of the matter is that very few pets are killed in home fires—just humans trying to rescue them. Animals generally sense danger long before humans and will seek out the safest spot. Also, where is the clearest air? Near the floor, the same place your pet is. A dog or cat often slips unnoticed through an open door at the same time its owners run out of the house. Its first inclination is to hide and not come out until all the excitement has died down. Even if your pet is still in the house, chances are the fire fighters will save him.

A while back, there was a very serious residential-hotel fire in Chicago that claimed several lives. At least two of those people died needlessly. The man who discovered the fire alerted his next-door neighbor, who was a good friend of his.

1-4 Don't go back into a burning house to save your pet—that's the fire fighters' job.

Both men started out of the building. Then the neighbor decided to warn a lady down the hall. The lady started out, too, but then she went back to get her purse and the neighbor waited for her. Neither one made it. How much could have been in that purse? Ten dollars? A thousand? No matter how much it was, it wasn't worth dying for. The same goes for priceless works of art or for Coney Island mementos. They're just objects, and people are more important than objects.

Vanity has often been a killer, too. Many people have refused to leave a burning house without getting dressed or finding their wigs or putting in their teeth. If there is a blanket or coat handy, grab it. If not, get out as you are: wearing a frilly nightgown, bald and toothless, or stark naked. It's a lot easier to get over a little cold air and embarrassment than to survive smoke inhalation or burns.

It can't be said too often: Never reenter a burning building. Wait for the firemen to arrive. They know how to fight a fire and how to get into a burning building to look for anyone trapped inside. They have the proper equipment to protect themselves and those they go in to save. In short, they know what they're doing—it's their job; they've been trained for it. You can help them best by telling them any important information they need to know and then staying out of their way.

2. WHEN THERE'S NO WAY OUT

There is the possibility that, try as you will, you cannot find a way to get out of a burning building. You might find every door blocked by flames, or encounter smoke so thick that you can't get through. If this is the case, and you can't leave the room you're in, don't give up. You can still save yourself.

If There Is No Way Out . . .

Seal doors and ducts.
Use any available water to dampen the seal.
If the room is full of smoke and a window can be opened and reclosed, ventilate the room.
Stay low and cover your mouth and nose.
If there's a telephone in your room, call the fire department or a neighbor and let them know where you are.
Wait, keep calm, don't panic.

Your first concern is air to breathe. Smoke and fumes from a fire—any fire—are toxic and will poison the air. So if smoke is seeping into the room you're in, the first thing you have to do is seal off the leaks. Do this by stuffing anything available into the cracks around the door and into or over all heating and air-conditioning ducts. Use sheets, towels, clothing—anything you can find in the room that you can push into the cracks. Even paper will work—tear up a book or magazine. You might have to use a knife, key, pencil, or some other tool to get the stuffing in far enough. If you can find some water to dampen the stuffing, so much the better. Look around. A vase of flowers or a tropical fish tank might contain enough water to make the stuffing damp and thus slow down the fire.

Doing nothing else besides sealing the door between you and the fire can improve your survival chances. The important thing, if you are unable to escape the building, is to buy time. A door—even a lightweight, hollow-core bedroom door —can prevent a fire from getting at you for as long as forty-five minutes. A fire will generally take the easiest path—that is, it will burn along a route with the best supply of fuel and oxygen, usually a fairly straight line. It will not turn a corner unless something draws it in that direction. By closing and sealing your door, you lessen the chance of the fire turning in your direction. I have seen bedrooms totally untouched by a fire that completely gutted the hallway outside but didn't even blister the paint on the inside of the door.

Suppose you have blocked off the source of smoke, but the smoke already in the room is so heavy you can hardly breathe. You can't leave the room—there is no escape. Under these conditions, *if and only if* a window can be opened and reclosed, you can ventilate the room for a short time. With double-hung windows, pull the top section down six inches or so to allow the gases that are gathering at ceiling level to flow out. Smoke and fumes are lighter than air and float upward. When they accumulate, it's as if the room were being flooded —but upside down. Opening the top of the window permits the smoke to form a kind of reverse waterfall, emptying noxious fumes up and out of the room. For a breath of fresh air under unbearable conditions when you can't escape, open the bottom of a reclosable window about eighteen inches and stick your head and shoulders outside. You might have to tent your head with a blanket or drapery so that the smoke-laden air can flow out over your head while you breathe the fresh air below. Cover your mouth and nose with a cloth to filter out as much smoke as possible. Again, it's more effective if you have dampened the cloth.

Of prime importance is to let someone know where you are so that you can be rescued as soon as possible. Surprisingly, many people will sit right next to a telephone and never think

-1 To slow entry of smoke, stuff material into cracks around door.

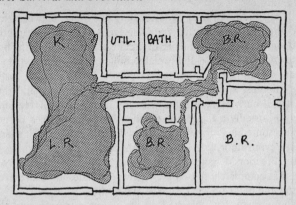

2-2 A traveling fire will bypass closed doors.

of using it to call for help. If there's a phone in your room, call a neighbor, or call the fire department and tell them exactly where you are—they can radio your location to the fire fighters outside your house.

Now comes the hard part: waiting. Seconds can seem like hours, and minutes like days while you are crouched at a window waiting to be rescued. You have to keep telling yourself to stay calm and alert, because any moment could bring a new crisis that you will have to deal with. You can do it! The secret is to be prepared, to know what is happening around you, and to know that you can handle it.

The enemy at this time is not just the nearby fire and smoke but also your own fear. You can't give way to it. By repeatedly going over in your mind all the information you've acquired and the steps you have previously practiced, and by constantly reevaluating your situation, you can keep yourself from panicking. Look at it this way: Any death in a fire is a premature death. There is no reason why you or any member of your family should perish in a fire. Thorough knowledge, preparation, and practice are your best defense against becoming a casualty.

2-3 Stay calm as you wait to be rescued.

3. THE NATURE
OF THE BEAST

Once a fire begins, it takes on the characteristics of an extremely clever and dangerous predator:

A FIRE FIGHTS TENACIOUSLY FOR LIFE. The tiniest spark seems determined to live. It will hide from you while it builds and gathers strength. You may not be able to see it, smell it, or feel it, but it's there, growing and getting stronger.

A FIRE HAS A RAVENOUS APPETITE. It will consume material that you never would have guessed would burn. That sturdy-looking wall in your family room? The whole thing could burst into flame and be devoured within minutes. When it's getting started, a fire will burn dustballs under furniture, paint off brick walls, or sealer off concrete floors. The more it eats, the more it wants. Fire is a chemical process. As combustible material is heated, a reaction takes place in which gases are released into the air. These gases combine into other combustible material—the fire is burning pre-existing fuel and creating additional fuel at the same time.

A FIRE STEALS OXYGEN. A fire must have oxygen to burn, and it uses up that oxygen at a very rapid rate. If it is to continue, its supply must constantly be replenished. The expression "It burned *up*" is literally true. Heated air rises. As a fire uses more and more fuel and oxygen, it gets increasingly hot, heating all the air around it. As the heated air moves upward, cool air (a fresh supply of oxygen) rushes in near the base of the fire to replace it. Result: a kind of windstorm that creates the type of environment in which a fire can flourish and grow.

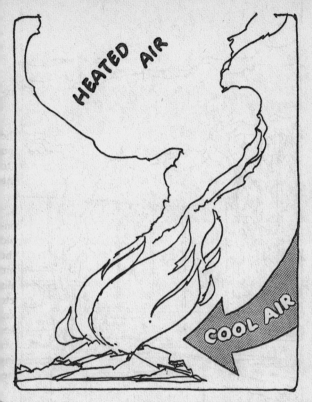

3-1 As heated air rises, cool air rushes in, bringing a fresh supply of oxygen.

A FIRE SOON BECOMES AN UNCONTROLLABLE MONSTER. Within seconds, a seemingly small fire can engulf an entire room in blinding flame. The heated air that a fire generates gathers at the ceiling level. Once enough of that superheated air collects—and this can happen within minutes from even a small wastebasket fire or a rangetop grease fire—it raises the temperature in the room to the flashover point. At that moment, everything that is combustible in the room will instantaneously burst into flame.

A FIRE WILL FOLLOW THE PATH OF LEAST RESIS-TANCE. Just as a predator will pursue the weakest animal in a herd, a fire will go after the easiest fuel it can find. It will rage down hallways ignoring rooms with closed doors, or it will jump across hard-to-burn objects to ignite a pair of flimsy curtains.

A FIRE IS NOT TO BE TRUSTED. A fire will double back and return to an already burned area looking for missed fuel. It will "play dead," letting you think it has burned itself out, and then spring back to life. It will suddenly release tremendous energy, exploding out of its confines into new areas. Only professional fire fighters can make the determination that a fire is really out. Never trust your own judgment; that little fire you think is dead could suddenly revive.

I remember a man whose cigarette ignited his mattress. He awoke to find smoke in the room and a small smoldering flame beside him. He quickly yanked the mattress from the bed and dragged it into the bathroom, where he ran water into the burn hole. Then he returned to his bedroom and went back to sleep in a chair. It was his last mistake. That tiny smoldering mattress fire had tunneled deeply through the mattress clear to the other end. Since the man had soaked only about half the mattress, the fire had the entire other half of the mattress to burn, plus a shower curtain. The fumes that fire gave off were silent, odorless, and deadly, and the man died without ever knowing what killed him.

A FIRE SENDS OUT SILENT KILLERS. Often long before you can see any flames, hear a crackling sound, or smell any smoke, a fire may already be sending out extremely poisonous gases. It's as if the fire were trying to neutralize any enemies capable of destroying it by destroying them first. These gases kill quickly, silently, and invisibly. One of the

2-2 Fire will jump across hard-to-burn objects and ignite something more flammable.

gases, carbon monoxide, is lighter than air, clear, and totally odorless. This means that a small fire from a cigarette dropped on the carpet behind a chair or a smoldering pile of rags in the basement can fill an entire house with enough lethal gas to kill everyone sleeping upstairs. And there may be only a couple of hundred dollars' actual property damage.

A FIRE DOESN'T KNOW THE MEANING OF "FIRE-PROOF." Have you ever stayed in a "fireproof" hotel? Most hotels are, nowadays. Even the M-G-M Grand Hotel in Las Vegas is "fireproof." But that didn't prevent eighty-four people from dying in it. "Fireproof" only means the building won't burn; it doesn't say a thing about the contents. The building I live in is a good example: It's constructed of reinforced concrete with firebreak dividing walls and self-closing steel corridor doors. But my home is filled with all kinds of things that would burn—wood and plastic furniture, clothing and curtains, air-conditioners and television sets. My entire home could go up in smoke, and my downstairs neighbor might not even know it. The only thing the building manager would have to do is clean out the debris, replace the wiring, and apply a fresh coat of paint, and the place would be ready to rent to the next occupant.

A FIRE DIES ONLY OF SUFFOCATION OR STARVATION. Once it has started, the only ways to kill a fire are to remove its food supply (fuel), smother it by denying it oxygen, or cool the material below its ignition point. These things take a determined effort and can best be done by trained and equipped professional fire fighters. For anyone else to decide a fire is dead and no longer a threat is just plain stupid. The fire department should always be called, even though you may think you have handled the problem. That's what they get paid for.

3-3 Inset in upper left shows first and second floors of house filled with smoke and carbon monoxide from small rag and trash fire in basement.

Classes of Fires

△ **Class A**   Trash, wood, paper, cloth, some plastics; generally dry, nonelectrical material

▣ **Class B**   Grease, flammable liquids

© **Class C**   Electric motors and equipment

Mixed Class   Many, if not most, fires fall into this category—that is, they involve combustibles from two or three classes. Because of that, they present complex problems.

Let's take a look at different kinds of fires and what makes each of them burn. It's essential to know just what you are fighting in order to decide whether or not to extinguish a fire. Common fires (and by this I include public building, house, and vehicle fires) can be grouped into three classes, as shown in the preceding box.

△ *CLASS A FIRES,* or other kinds of fires involving Class A materials, are the most common types of home fires. They are caused by such things as cigarettes accidentally dropped into wastebaskets or between the cushions of an upholstered chair or sofa; kids playing with matches; a tipped-over candle on a carpet; space heaters sitting too close to wood, plastic, or upholstered furniture; a curtain or dish towel hanging too close to a range burner; sparks from outdoor cooking and indoor fireplaces—the list is endless.

Once a Class A fire has passed the smoldering stage and begins to flame, it will often move very, very fast, and quickly generate a tremendous amount of heat. Something as simple as a drapery brushing against a lighted candle can ignite and flame so rapidly that there is absolutely no time to pull it

3-4 A Class A fire spreads fast. Get out immediately.

down or attempt to extinguish it. Yell a warning, get out immediately, and call the fire department. The twenty seconds it takes you to run into the next room for a fire extinguisher may be all the time needed for the flames to spread. Never waste time trying to be an amateur fire fighter or a hero. You may make things worse and lose your life in the bargain.

B *CLASS B FIRES* are commonly those involving flaming grease in a cooking utensil, or petroleum-product liquid solvents, cleaners, and paints that have vaporized.

Probably most people have seen a skillet of bacon or a broiler with fat drippings suddenly burst into flame. It isn't unusual, but it is dangerous. Quick thinking can often prevent a cooking fire from getting out of hand. Keep a lid the size of your frying pan handy while you fry bacon or pork chops; you can slide the lid over the pan if the grease suddenly flames up. If you're broiling a fatty piece of meat and the grease catches fire, close the broiler and oven doors and turn off the heat. Either of these methods is usually effective in smothering a small grease fire. The thing to do, of course, is to be watchful while you're cooking. Don't allow the skillet to get too hot or let too much grease accumulate. Use a broiler with a slotted top tray or cover the pan with a sheet of aluminum foil perforated liberally (use the tines of a fork) so grease will not be exposed directly to flame. *Never attempt to move a flaming skillet or pan*—spilled flaming grease will spread the fire. *Never throw water on a grease fire*—water thrown into flaming grease will cause an explosion, spreading the fire.

Solvents and cleaning solutions generally burn by vaporizing—that is, they evaporate and the gas given off explodes and burns. As with grease, water should never be used on any flaming liquid, because it will only spread the fire. Some va-

3-5 Three common causes of electrical fires: an overloaded outlet, a frayed cord, and a lightweight extension cord used with a heat-producing appliance.

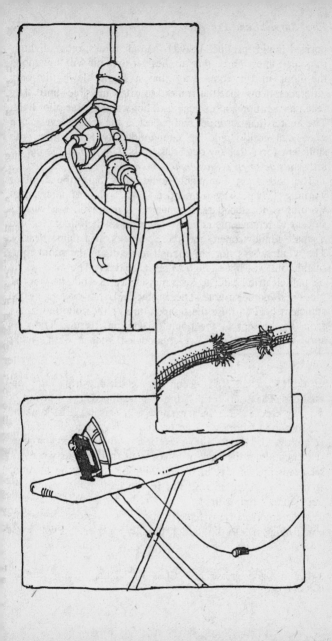

porized liquids produce heavier-than-air fumes, others lighter-than-air fumes. Those that are heavier than air will flow along the floor in the same way that a liquid flows—the only difference being that the fumes are often invisible—until they reach an ignition source such as a hot-water-heater pilot light. The lighter-than-air gases will collect at the ceiling level. Such gases can easily build up enough density to explode with sufficient force to blow out brick walls. *Use flammable liquids only outside or in a thoroughly ventilated area.*

The labels on flammable-liquid cans bear the familiar warning "Keep away from open flame." Unfortunately, that warning is not adequate. I remember a man who was using a solvent to remove tile adhesive from his kitchen floor. He had a small window open, and there was no open flame nearby. The window was not big enough to ventilate the room thoroughly but was big enough to let in cold air. The man set up a small electric heater to keep himself warm while he worked. No open flame, right? The heater did, however, give off sufficient heat to raise the temperature of the collected fumes from the solvent to the point of ignition. Result: The man survived—barely. His newly remodeled kitchen didn't make it.

© *CLASS C FIRES* occur in electrical wiring and appliances. They can result from overheating (e.g., a string or drapery cord catches in a window fan, causing it to jam and the motor to labor) or from broken or frayed cord insulation (e.g., the dried-out rubber covering on an extension cord stretched along the back of a basement shelf breaks when you set a corrugated box of old clothing on it, and the resulting spark starts a smoldering fire in the carton that eventually bursts into a major fire).

A common cause of electrical fires is overload—too many cords plugged into one outlet. Usually an overload will cause

3-6 A stovetop grease fire (Class B) igniting nearby curtains (Class A).

the fuse to blow, but too often people ignore this warning and place a penny behind the fuse, enabling them to continue using the badly overloaded circuit. Eliminating that measure of protection is flirting with disaster.

Another kind of overload results from using a lightweight extension cord for a heat-producing appliance such as an iron or toaster. Many such extension cords have plastic connections at each end that, when heated by the amount of current required by the appliance, can melt or burn, causing a serious fire. Something else to avoid is the use of a light bulb with a greater wattage than recommended for the fixture. Replacing a forty-watt bulb with a hundred-watt bulb might cause enough additional heat to damage the receptacle and cause a fire.

When an electrical fire occurs, your first action should be to disconnect whatever has caused the fire or else to turn off the power at the fuse box or circuit-breaker box. If that isn't possible, leave immediately and call the fire department. *Never put water on an electrical fire.* Electrical current can travel along a line of moisture and may electrocute you if you happen to be standing on a damp spot.

Electricity can also travel throughout a metal cabinet (including the "steel box" of a refrigerator, cooking stove, air-conditioner, or heater), causing any combustible coming into contact with it to burst into flame. As an example, an appliance with a faulty cord sitting on a steel kitchen sink/cupboard unit can electrify the entire cabinet, burning you and anything else that touches it.

MIXED-CLASS FIRES involve more than one class of combustibles. For instance, a stovetop grease fire (Class B) may ignite nearby curtains, dish towels, or potholders (Class A); a discarded cigarette (Class A) could burn through the insulation of a lamp cord (Class C); spilled lawn-mower gasoline (Class B) could be ignited by the mower's hot motor and set fire to a pile of scrap lumber (Class A); or a spark from an electric saw (Class C) could jump to a pile of sawdust (Class A) or an open can of paint thinner (Class B).

The best thing to do when any fires like this get started is to get out fast, yell to warn anyone else in the house, and call the fire department. Smothering might work on burning liquids, maybe even on a small pile of trash, but if there is electricity involved, it won't help. And a bucket of water thrown at a flaming wastebasket could also knock over a waste-grease jar and cause an even larger fire. Don't take the chance of being wrong. Get out, close the door on the fire, warn others, and let the firemen handle it.

4. EARLY WARNING SYSTEMS

So far I've talked about *seeing* flames or *smelling* smoke as the way in which you're likely to discover a fire burning in your home. However, if your home is properly protected, your first warning of danger is probably going to be the buzzer you hear when your smoke detector senses smoke particles in the air.

When your smoke detector sounds an alarm, believe it! Don't think it's just a false alarm. Smoke detectors are generally very reliable and don't often go off spontaneously. Even if a false alarm should occur, it's better to be safe than sorry.

Remember: Smoke detectors are early warning devices. *They do absolutely nothing to put out fires!* They do not remove the danger—you must remove yourself from the danger when they warn you.

A smoke detector is not expensive—it is one of the best safety investments you can make. Recent data show that without a detector, the risk of dying in a fire is almost twice as great as that if you have one.

There are different kinds of smoke-detection devices, and they work in different ways. It's important that you know those differences.

Kinds of Fire-Detection Devices

Ionization smoke detectors	Systems that combine
Photoelectric smoke detectors	two or more of these
Heat detectors	methods of detection

IONIZATION SMOKE DETECTORS contain a tiny amount of radioactive material within a small chamber. The radioactive material causes the air within and moving through the chamber to become ionized, or electrically conductive. Electric current then continuously flows between two electrodes in the chamber. If smoke particles enter the chamber, the flow of electricity is decreased and a warning buzzer sounds.

The amount of smoke necessary to impede the flow of electric current through an ionization detector is extremely small. In fact, the smoke particles may be so tiny that they are invisible to the human eye. Enough of them, though, can decrease the current flow sufficiently to make the detector go off in a matter of minutes—before you can either see or smell any evidence of fire. This sensitivity is especially useful in situations where hot blazing fires can occur. Burning paper and some wood fires are examples of fires that can burn very fast and efficiently, create a lot of heat, spread rapidly, and give off a relatively small amount of smoke.

Ionization detectors are able to detect some of the most le-

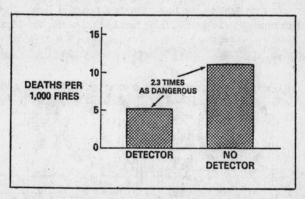

4-1 Life-saving effectiveness of smoke detectors in one- and two-family homes. (From *Preview Residential Fires in the United States 1979*. Federal Emergency Management Agency/U.S. Fire Administration: NFIRS, 1977–79.)

thal of all products of combustion. If you were awake when your house began to fill with smoke, you might begin to feel light-headed or sick to your stomach. But if a fire started while you were asleep, chances are the smoke would not disturb you, and you would never wake up. An ionization detector protects you and your family against that happening.

Some people are reluctant to purchase ionization detectors because they fear that radiation will escape. There are no detectors on the market that contain enough radioactive material to cause concern. Ionization detectors have been carefully tested and found to present no significant radiation risk. In fact, the Nuclear Regulatory Commission says, "If you held an ionization smoke detector close to you for eight hours a day through a whole year, you would receive only a tenth as much radiation as you'd get on one round-trip airline flight across the USA."

PHOTOELECTRIC SMOKE DETECTORS work very differently. Using a beam of light, they are able to "see" smoke in the air. Like the ionization detector, they have a chamber through which air constantly flows. Inside the chamber is a tiny light source that projects a beam of light across the chamber. As long as nothing breaks that beam, the detec-

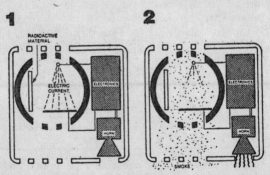

4-2 Schematic drawing of an ionization smoke detector. (Courtesy of U.S. Consumer Product Safety Commission.)

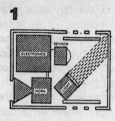
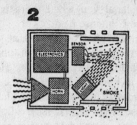

4-3 Schematic drawing of a photoelectric smoke detector. (Courtesy of U.S. Consumer Product Safety Commission.)

or remains quietly on duty. But if enough smoke enters the chamber to deflect the beam to a sensor, or "electric eye," a buzzer will sound.

Some slow-burning fires, such as smoldering upholstery, generate a fair amount of smoke that can be "seen" as it passes through the photoelectric detector's air chamber. Fires of this type produce smoke particles that are larger than those of hot, clean-burning fires and so are more likely to break the light beam. Once the beam is broken and enough light is deflected to the sensor, electricity is generated that sounds the buzzer.

HEAT DETECTORS are useful for home protection when they are properly placed. Remember, it is almost always smoke inhalation (asphyxiation) that kills people in home fires. If you have a heat detector outside your bedroom door, it isn't going to go off until a considerable amount of heat has actually reached that spot. By that time, there would probably be so much smoke and poisonous gas that a heat detector's warning would be too late to save you.

Heat detectors work in much the same way as a furnace thermostat. They contain a special kind of metal that will bend when heated, closing a switch and causing a buzzer to sound. Some contain a small piece of soft metal that, like an electrical fuse, will melt when it gets hot enough. Whichever kind you choose, you have to remember that the detector will *not* signal you when there is smoke in the air, only when there is intense heat.

A heat detector placed in a furnace room, over a clothes dryer, in an attic, or in an attached garage can provide lifesaving added protection. Heat detectors can also be effective in kitchens, where smoke detectors are not recommended because of the number of alarms triggered by burning food. Those kitchens that have adjoining laundry rooms or furnace and hot-water-heater utility rooms should have heat detectors.

To assure maximum protection for you and your family, you need a combination of all three kinds of detectors. There are smoke detectors on the market that contain both ionization and photoelectric capabilities within the same unit. Some smoke detectors also contain lights that will turn on at the same time the buzzer sounds. They can help light your way to an exit and are particularly useful if the power fails for any reason (during a storm, for example, or if a fire has knocked out your home's electricity).

In short, no one detector is best for every location. Used in combination, though, different kinds of warning devices placed in strategic areas of your home will go a long way toward saving your life and the lives of your family should fire strike.

When installing smoke detectors, be sure to place at least one between the most likely fire area and the sleeping area. Either a photoelectric or an ionization unit will protect you, but because of the different ways in which they work, it's best to have one of each. Two or more units offer more than double protection: It's unlikely that both would malfunction at the same time (assuming you keep each supplied with fresh batteries).

In general, it's wise to place a photoelectric detector in the living area of the home, either downstairs in a two-story house or at the entrance of a hallway leading to the bedrooms in a single-floor home. That way the detector is most likely to sense the large smoke particles given off by a smoldering fire in upholstery. If there are any smokers in your household,

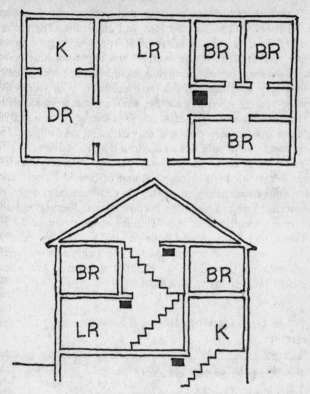

4-4 Floor plans showing placement of smoke detectors. Above: single-level dwelling; below: multistory dwelling.

there should be a photoelectric detector in each of their bedrooms.

Place an ionization detector just outside the bedrooms so that it can warn you of fast-spreading, high-heat, low-smoke fires. Another detector just outside the kitchen is a good idea, too. Don't put it *in* the kitchen, though, where it might get clogged with grease or sound false alarms while you're cooking a meal.

There's overwhelming evidence that you're safer inside a closed room with a door between you and a fire. Therefore, you and your family are better off if every bedroom door is kept closed at night. If you're afraid you might not hear your smoke-detector buzzer through a closed door, select a unit with a very loud horn. If you or someone in your family is hard-of-hearing, then put another smoke detector inside that particular bedroom. The safest thing, of course, is to install a detector inside every bedroom as well as in the hallway. In most cases you're only talking about a hundred dollars.

The biggest objection you may have to closing doors at night is that it's hard to hear children or invalids. There are intercom systems designed so that they are activated when a baby cries. You can install an in-room smoke detector and an intercom system for less than $200. It's money well spent.

Don't think I'm trying to make you feel guilty. I'm actually trying to relieve you of guilt, the terrible guilt you would feel if something happened to your children and you hadn't taken the right precautions to save them.

What to Look for When Buying a Smoke Detector

Power supply: Battery is easiest

Loud buzzer or horn

Easy installation

Test button

Low-battery warning—a must

When shopping for a smoke detector, don't buy by price alone. Check the features in the box before making your selection.

All smoke detectors are electric, but they may be powered in three different ways: wired into the electrical system of the house, plugged into a regular wall outlet, or operated by batteries. Considering all factors, batteries are cheaper and easier to use.

Connecting a detector permanently into the house current is not terribly complicated, but it does require at least rudimentary knowledge of wiring procedures. If you don't know how to do it, you have to find someone who does. Either way, it's an expense and an unnecessary delay. The advantage, of course, is that you never have to worry about dead batteries. It is unlikely that a fire would knock out the electricity before the detector sounded its warning.

Tying your smoke detector in with a home burglar alarm may seem like a good idea, but don't do it. The two security systems should have very different warning alarms. You don't want to have your children hide, thinking there's a burglar in the house, when they should be getting out and away from a fire.

Some older photoelectric detectors have wires that plug into ordinary wall outlets. These units contain small light bulbs, like flashlight bulbs, that remain lighted all the time. Because they are always on, they draw too much power for batteries to be used and so must be plugged into the electrical system. They work fine and are inexpensive to operate. If you use one, however, install a special device over the plug to insure that it stays connected to the outlet.

Newer photoelectric units contain light-emitting diode (LED) light sources that flash on and off intermittently and thus use very little electricity, making it practical for them now to be battery-powered. This feature also eliminates what some people may have considered an unsightly drawback, namely, a dangling cord.

Since both photoelectric and ionization detectors require so little power to operate, the cost of batteries is not prohibitive—one long-life battery will probably last several months, perhaps even a year. A word of caution, however: Most detectors on the market take inexpensive, commonly available batteries—beware of those that don't. I remember a recent apartment-building fire that took seven lives. It was caused by someone barbecuing on the back stairs, and so one of the

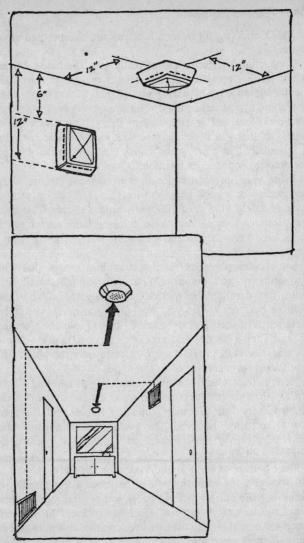

4-5 Distance from wall-ceiling corner for both ceiling and wall installations.

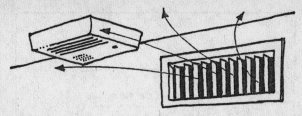

4-6 NEVER install a smoke detector in front of an air supply duct.

exits was unusable. But that wasn't the worst part. The building was well maintained and had adequately placed smoke detectors on every floor—and not one of them worked. Why? They required a unique kind of battery that wasn't locally available and cost over $10. There might just as well not have been a single detector in the entire building. The batteries were dead and couldn't be replaced. So were the unlucky seven.

When shopping for a smoke detector, don't take the word of the advertising copy on the package. Ask to take it home and test it. Put in a battery and use a cigarette or snuffed candle to make it buzz. Is it loud enough? Take it back if you don't think so. Remember, you'll probably have to hear it through a closed door. You should select a unit with a harsh, unpleasant alarm, either pulsing or steady. Buy one that is really going to alarm you.

Most detectors are easy to install. Follow the directions on the box. In general, a detector should be hung either on the ceiling or on the wall. If you're putting it in the hall, the best spot is right in the middle of the ceiling. Whether you choose the ceiling or the wall, don't hang it in a "dead air" area, or closer than six inches to the wall-ceiling corner. Since smoke rises, never hang a detector lower than about twelve inches from the ceiling. And don't install your detector near a ceiling-level heating or air-conditioning duct—the forced air will prevent smoke from remaining in the air chamber long enough for the detector to sense it.

4-7 Snuffed-candle test.

If you select a battery-operated model, you can have your detector installed and protecting you within five minutes after you bring it home. With excellent smoke detectors available for just a few dollars and so easy to install, how can anyone afford *not* to have one?

Most units have some sort of provision for testing the alarm to make sure it's on guard and working. Usually it's a little button somewhere. By pushing it, you can make sure the battery is still working and that the unit is ready to sound an alarm if needed. Do this once a week. An even better method than the test button, though, is the snuffed-candle test. This really simulates the kind of situation your detector is guarding you against. To test your unit this way, light a candle and let it burn for a minute or so. Then blow it out and hold it a few inches beneath the smoke detector so that the rising smoke can enter the detection chamber. Don't be impatient; it may take thirty seconds or so before the buzzer sounds. You can stop the alarm by fanning the smoke away from the detector. To repeat: Test your unit once a week.

All good smoke detectors have some sort of warning device that will signal you when the battery is low and ready to be replaced; if they don't have one, they're *worthless!* Some detectors begin a periodic beeping when the battery starts to run down; one model has a small red flag that appears when it's time for a change. But don't wait until your detector tells you it needs a new battery before you think about running down to the drugstore. Keep extra batteries on hand (bulbs, too, if you have a unit that uses one) and replace dead ones right away. Never, never borrow a battery from a smoke detector in order to operate some other appliance or toy. There's nothing you need to have working that's more important than a smoke detector!

Like any appliance, detectors need proper care to function properly. Dirt in the air, especially greasy dirt in a kitchen, will clog a smoke detector and make it ineffective. Periodically (three or four times a year) take the cover off your detector and clean it with a vacuum cleaner brush. Some units require additional care and cleaning—follow the instructions on the carton.

I can't stress enough the importance of taking care of your smoke detector. If you're not willing to keep it clean and supplied with a live battery, then you might just as well hang a nice, bright, shiny hubcap on your wall and call it a smoke detector. You're only fooling yourself.

TIP

Consider giving a smoke detector as a Christmas or wedding gift. It's not only a good practical gift, but it tells the people you give it to that you really love them and care about them.

Home smoke and heat detectors are warning devices only. Large public buildings such as department stores, exposition halls, schools, and hospitals may have other automatic fire-fighting equipment as well as smoke detectors. (Unfortunately, far too few large department buildings and hotels have installed this type of protection.)

SPRINKLER SYSTEMS. This is absolutely the best kind of automatic fire-fighting system that you can install. People are beginning to put such systems into their homes now that sprinkler heads are more attractive and will not only turn *on* during a fire but turn *off* again once the fire has been extinguished. Some will even turn back on again if the fire should happen to double back into the same area. This greater control reduces the possibility of extensive water damage. Sprinkler systems are not nearly as expensive as they used to be, especially with the advent of lightweight plastic pipe.

VENTILATOR CUTOFFS. Smoke detectors installed in ventilator shafts will automatically close off sections of ducting when gases are present. This reduces the danger of poisonous smoke and fumes being distributed throughout an entire building.

FIRE DOORS. Like ventilator cutoffs, steel compartmentalizing doors can be set to close in response to either smoke or heat detectors, thereby confining fire to a relatively small area.

OTHER HOME ALARMS. Some home smoke and/or heat detectors can be rigged to sound loud outdoor sirens that will alert neighbors to an emergency so that they can notify the police or fire department. In addition to a siren or horn, you might want to consider an outside speaker that would continuously broadcast a very loud recorded message such as "There is a fire emergency in this house. Please call the fire department at once!" Gadgets like this can be very helpful if the house occupants are handicapped or elderly. One of the newest types of alarm available in a few areas is a two-way cable-television system that can sense heat, smoke, or intruders in a home and automatically summon assistance. It monitors the home every ten seconds, twenty-four hours a day, whether anyone is at home or not and whether or not the television set is turned on. A more conventional direct alarm system wires a home's heat and smoke detectors into a central dispatching switchboard.

5. FIGHT OR FLIGHT?

Suppose you're ironing and the telephone rings. You're expecting an important call and you rush to answer it. Apparently you haven't set the iron down quite as firmly as you normally do, for when you return, you find that the iron has toppled over, burned through the blouse you were pressing, and has started a fire. It's just a small fire—a bit of smoke, possibly a little flame. Do you try to put it out? Or do you run out the nearest door? Depending upon the extent of the fire, the resources you have at hand, and, most important, your own ability to act fast, either choice might be correct.

Think! Before You Try to Fight a Fire, You Must Decide:

Do you have an escape route?
Is the fire containable?
Do you have the means to fight the fire?
Will your fire fighting be a delay that will allow the fire to spread?
Are you endangering the lives of others?

Having discussed the procedures for getting out of a burning building, let's take a look at some of the considerations and judgments to be made in deciding whether or not to fight a small fire.

Never take the time to fight a fire at the risk of having your exit blocked. This applies even to what you might consider small, insignificant fires such as the one described above. If you were to enter your kitchen, for example, to find a fire between you and the back door, your best course of action would be to turn around and get out *immediately* through an other exit. You already know that if you are unable to extinguish the fire, you can't use the back door. Don't risk letting the fire get bigger, filling the air with so much heat and smoke that you won't be able to make it to the front door either. This is especially important in a house where the exits are a considerable distance apart, and it is crucial in a long "railroad flat" (an apartment laid out with all the rooms in a straight line like the cars of a train) where the only other exit might be down a long hall.

Superheated air can travel faster than you can run and could cause a flashover that would instantaneously ignite everything along your escape route, including the door you hope to reach. In a situation like this, your only choice is to yell a warning to anyone else at home, close a door on the fire if possible, and get out fast. Don't try to telephone the fire department until you are safely out of the building. The precious seconds it takes for your call to get through could be just long enough for poisonous fumes to collect between you and your only other means of escape.

Always keep your back to the door while fighting any kind of fire with any kind of equipment. So positioned, you can quickly turn around, leave the room, and close the door behind you if you see the fire even *begin* to spread. Never allow a fire to move around behind you—get out before it has the chance.

Deciding if a fire is extinguishable is a subjective judgment that you have to make immediately. While you stand staring and wondering whether to leave or fight it, the fire is going to be growing, gathering strength, and giving off lethal gases. The only thing you should consider is whether or not the fire is containable—that is, whether you can put it out before it spreads or before it reaches new fuel. If you have any doubt, your question has been answered. Get out!

That ironing-board fire is probably one to get out and away from. Unless you were able to disconnect the iron and douse the fire with water or smother it with an extinguisher—without further endangering yourself by knocking over the board, burning yourself on the hot iron, or giving up your clear path to an exit—the chances are you would only cause a needless delay in summoning the fire department and perhaps place yourself in greater danger.

A grease fire in a skillet is a good example of the kind of fire you might be able to handle before it gets completely out of hand. By sliding a lid over the skillet and turning off the heat under it, you will smother the flame. Again, the most common mistake made in a situation like this is to try to move the skillet to the sink—by doing so, you could spill flaming grease that would spread the fire around the room or ignite your clothing. If you can't smother the flames with a lid or an extinguisher, get away from it and call the fire department.

Kitchen fires spread very quickly because there is often a thin coating—too thin to see—of grease from previous cooking covering the walls, ceiling, and other surfaces. The grease acts as an accelerant, helping the fire to grow.

Another kind of fire that is fairly easy to extinguish is a trash-can fire. If you can put a lid on it—a flat object such as a cookie sheet or board can be used—you may be able to smother it before it becomes large enough to spread. Despite the fact that a board itself is flammable, if it fits tightly enough over the top of a can or wastebasket, it can smother a small fire before getting sufficiently hot to burn. Once you have extinguished a fire this way, carry the entire can, with the "lid," outside, away from the house, and soak everything —contents and "lid"—thoroughly with water.

Equally important to consider, if you decide to fight a fire, is the means you have conveniently at hand to fight it with. An all-purpose fire extinguisher is best, of course, but there are other things you can use. Water is good for nonelectric, Class A fires—if it is handy. Running two or three rooms away and waiting while a pan fills from a faucet and then running back to the fire takes too long if you're faced with anything larger than a burning ashtray or a very small metal wastebasket fire.

A cooking fire can be extinguished with common house-

1 Leave yourself an escape route.

5-2 If you decide to fight a fire with an extinguisher, stand with your back to the door.

hold baking soda, and it's wise to keep a couple of open boxes near (not over) the kitchen stove. Baking soda is very effective in smothering flames. Not effective at all is flour or sugar. In fact, throwing either of these kitchen staples on a fire can be disastrous—flour creates a very fine dust that can explode with a devastating impact, and sugar will burn with a stubborn, hard-to-extinguish flame that can spread the fire.

A heavy piece of fabric can smother a small fire before getting hot enough to burn itself. Blankets, tablecloths (not plastic), draperies, rugs, and coats can be used. Try this only on a small fire, though, or you may find yourself actually holding fire in your hands. Smothering a fire means just that, depriving it of air. Don't use a blanket to beat a fire; by doing so, you may only spread sparks that will ignite other things in the room. (Never beat a flaming liquid fire with a blanket—you will only replenish the oxygen supply, and the blanket will pick up the fire and spread it everywhere you wave it.) Drop one or two blankets over the fire, and then, if there is no electricity involved, pour water over the whole thing. When you're sure the fire is out, carry the whole mess outside, away from the house. Make sure the floor under the fire is not hot. If you have any doubts at all, call the fire department.

Any of these techniques takes a bit of time, even if your "fire-fighting equipment" is right at hand. If you have misjudged and are unable to prevent the fire from growing and spreading, you've wasted valuable moments that can decrease your chances of escaping or increase the amount of damage done before firemen arrive. I can't repeat too often: If you have *any doubt at all* about your ability to extinguish a fire totally, don't even attempt to fight it; instead, close a door on it, warn others, and call the fire department from outside.

Delays caused by futile attempts to put out a fire can endanger other people in the building with you. If you live in an apartment building, you have a responsibility to your neighbors just as you do to members of your own household. By playing fireman in a situation beyond your capabilities, you place all their lives in jeopardy. Turn the circumstances

5-3 Smother a grease fire in a skillet by sliding a lid over it.

around. Wouldn't you rather your neighbor didn't take chances with your life?

One special point needs to be mentioned here: children who start fires. Children who play with cigarette lighters, can

-4 A cookie sheet may serve as a lid for a trash-can fire.

dles, matches, or anything else that starts fires are a menace to all who live with them. I would hope that your family training sessions would eliminate this problem, but no one can ever be certain. There is a special danger here. A kid who starts a fire is going to be scared—not so much of the fire as of being blamed for it. When that happens, he or she may try to avoid punishment by hiding the fire or running away from it and not telling anyone of the danger. You must impress upon your children the absolute necessity of letting someone know immediately if there is any fire danger—whether or not they had anything to do with starting it. There have been far too many children injured or killed because they were afraid to report a fire. Don't let your kids suffer the same fate.

The examples given here are of small, manageable, containable, fightable, escapable fires. How can you tell ahead of time if your fire is all these things? You can't! The best advice remains—when fire breaks out, get away from it, warn others, and call the fire department.

After You Have Put Out a Fire . . .

Feel the fire site; take all remains of the fire outside.
If the fire site is too large to move, call the fire
    department.
Ventilate the entire area.

Let's assume you've had a small fire in your home and that you've been able to put it out yourself. Is the danger over? Very definitely no! You may feel you've done everything possible and necessary to extinguish the fire completely. But you're not done yet.

After you have extinguished a fire completely, feel it with your hands. Break the remains apart and make sure they're cold. If you have used water on the fire, all the combustible material should be damp clear through. Fires involving Class A materials may be knocked out with a foam or dry

5-5 If you have any doubt about your ability to fight a fire, get out immediately and call the fire department from a neighbor's house.

chemical extinguisher but should still be drenched with water afterward.

Even if everything feels cold and damp, move it all outside anyway. Take all the fire remains far away from any buildings; a second soaking with a garden hose is not overkill, only good sense.

Some pieces of furniture, such as mattresses, upholstered chairs, and sofas, may be too heavy or bulky to carry outside. If that's the case, *call the fire department!* Even if you've soaked the padding, felt carefully down inside a burn hole, or poured water behind a big, unmovable piece of furniture, don't take the chance of a tiny bit of fire lying dormant and then springing back to life an hour or two later. It can happen—*it has happened!* It's the fire department's job to make sure that fires are out. Don't trust your own judgment.

Call the fire department also if any electrical wiring has been damaged by the fire. If the fire was near an electrical outlet or a wall switch or light, there is the chance that the wiring inside the protective plate may have been damaged and could short out when you try to use it. Of course, if the fire involved a lamp or some electrical appliance, you should not attempt to use the item again until it has been properly repaired. Going back to the ironing-board example, if you have thrown water on any electrical appliance, it must not be plugged in again for any reason until it's been thoroughly checked by a qualified repair shop.

If you have thoroughly doused a little fire, removed the burned material, and made sure there is no chance of the fire returning, you have one final step to take. Open the windows and ventilate the house completely. Remember that every fire, no matter how small, gives off dangerous fumes and that those fumes rise and collect near the ceiling of the highest part of the house. There may not be a high enough concentration of gases to be fatal from a little fire, but you, or more likely your small children, could still become ill from breathing the polluted air. Set up a fan or two and make sure you empty the whole house of gases.

In this chapter I've set forth the considerations that go into making that critical decision when a small fire breaks out: Should you or should you not try to fight it yourself? If the decision is yes, you won't get very far in most cases if there's no fire extinguisher handy. The next chapter tells you what you need to know about fire extinguishers—what kind to buy, where to put them, and how to use them.

6. FIRE EXTINGUISHERS

Every year brings another crop of "necessities" to the market. We fill our homes and workplaces with electric coffee makers, humidifiers and dehumidifiers, and computerized this, that, and the other thing. To make everything pretty and clean, we can choose from dozens of paints and paint removers, coatings and strippers, lubricants and solvents. Every one of these items is a potential source of combustion, and the fires that can result from their misuse can be stubborn and deadly.

Just as there are three types of fire—plus combinations of these three—there are three basic types of fire extinguishers—plus some that will work against combinations of fire classes. For an extinguisher to be of value to you, you have to understand its capabilities and limitations. You won't always have time to read the label on an extinguisher before you point it at a fire.

This doesn't mean you have to be acquainted with the workings of every commercial extinguisher on the market. All you have to remember is your ABC's.

Types of Fire Extinguishers

⚠ **Class A** Contains water or weak water-chemical solution.

🄱 **Class B** Contains foam, dry chemicals, or CO_2 "snow."

© **Class C** Contains dry chemicals.

Class A, B, and C extinguishers correspond to Class A, B, and C fires. To help you identify the correct piece of equipment at a glance, a different geometric shape is used as a symbol for each class—a triangle for A, a square for B, and a circle for C. Some manufacturers also color-code the background of the class symbols—green for A, red for B, and blue for C. These symbols are sometimes used with pictorial symbols of the type of fire on which the extinguisher should be used.

Let's take the types of fire extinguishers one at a time.

△ *CLASS A EXTINGUISHERS* are intended for use on fires involving most dry combustibles, such as wood, some plastics, paper, and cloth. They contain either plain water or a weak chemical solution (usually water and sodium bicarbonate, "baking soda"). The bomb-shaped tank seen so often on the walls of public buildings is the most common example of Class A extinguisher. When overturned, a small vial of acid inside the tank mixes with the soda-water solution and creates a gas that forces the water out from the hose. It has the advantage of requiring only a simple procedure for operation: Turn it upside down and it works. There are no valves, safety mechanisms, or triggers to operate.

Other Class A extinguishers may have pump handles and function in the same way as a garden sprayer you might use to kill weeds. The tank is filled with either water or the same soda mixture as in the other extinguisher and is operated by aiming the hose at the fire and pumping the water out.

A Class A extinguisher has a limited application. (A bucket of water could be considered a Class A fire extinguisher.) It may put out a small trash fire, but unless you aim the stream carefully, you may find you have more fire than you do water. And, of course, once the water supply is exhausted, the extinguisher is of no further use. Before using this type of extinguisher, you must be positive there is nothing but Class A fire material involved—no wiring, no burning liquids, no grease, no plugged-in appliances. Otherwise you could compound the problem and further endanger yourself and others.

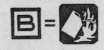

 CLASS B EXTINGUISHERS are designed for use on flammable liquids, greases, and similar materials. The most effective type contains a dry chemical that shoots out in a powdery cloud and smothers the fire. Carbon dioxide extinguishers can be used on Class B fires, but their advantage of leaving no residue to clean up is outweighed by their lesser effectiveness and their greater cost. Foam extinguishers work on small, kitchen grease fires, but they are too limited for most purposes.

A word of extreme caution here. A Class B extinguisher works fine on a grease fire or on a flaming solvent where the fire is on the surface, but a wood or paper fire's core will probably be deeper within the material. This means that the fire could flame up again if it works its way back out or tunnels to a new location. Therefore, it's imperative that Class A material be drenched with water once the initial blaze is controlled.

© *CLASS C EXTINGUISHERS* contain a dry chemical that is nonconductive and therefore effective on electrical fires. It, too, acts by smothering the fire. An extinguisher with only a Class C rating will not be particularly effective on fires involving wood, paper, or fabric for the same reason as above.

The A, B, and C classifications for different fire extinguishers are applied by both Underwriters Laboratories (UL) and the National Fire Protection Association

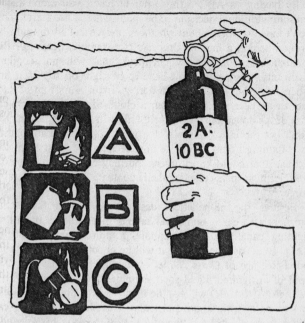

6-1 An ABC extinguisher.

(NFPA). Designations for effectiveness are made after extensive testing on various kinds of fires and are guides you can use when purchasing a home fire extinguisher. When shopping for extinguishers, you'll probably find that most carry multiple designations.

I recommend buying an extinguisher that is labeled for use against all three classes of fire—A, B, and C. That way you avoid making a serious and potentially dangerous mistake. An ABC extinguisher can be used against any fire you're likely to encounter in your home. But bear in mind that a Class A fire (deep in upholstery or a mattress, for instance) may still have a stubbornly smoldering core that requires further action.

By having an ABC extinguisher at hand, you aren't faced with an additional decision to be made when a fire starts. You don't have to worry about whether an electrical cord might be broken behind a flaming drapery, for example, because ABC extinguishers contain dry chemicals that will not conduct electricity. Also, you don't have to be right on top of a fire when using an ABC extinguisher; you can aim from several feet away because it sends out a cloud of powdery chemical that drops down like a lid over the fire. In fact, it's best if you don't get too close (eight to ten feet away is probably best) to a flaming liquid or grease fire, because the blast from the extinguisher could knock over a burning can or pan. Pressurized extinguishers release their contents with considerable force.

Besides the alphabetical designations, home fire extinguishers also carry numerical ratings that indicate their firefighting capability. For example, a unit might be labeled 1A:10BC. That means that it will be effective against a fairly good-size fire of Class A fuel or flaming grease as well as a spill of an ignited liquid such as paint thinner. A unit with a rating of 2A:10BC will handle fires twice as big. It's impossible to be specific about particular fires since conditions can differ so much, but a rating of 2A:10BC is probably adequate in most situations. Notice that there is no separate numerical rating for a Class C fire; that's because the "C" on the label simply means that the extinguisher contents are safe for use on electrical fires.

An important thing to remember is that an ABC extinguisher can be used to smother a person's burning clothing. The dry chemical contained in the extinguisher can be washed off. Immediate action is called for if someone's clothing is burning, and a quick blast from a home extinguisher could save that person's life.

Operating a Fire Extinguisher

Keep an exit available.
Remove safety mechanism.

Brace yourself; face fire.
Aim at base of fire.
Spray with sweeping motions.
Operate extinguisher with several short bursts.
If unable to extinguish fire, get out at once.
If fire extinguished, check to make sure.
Remove all burned material from building.
Ventilate.

Now that you know something about how home extinguishers work on fire, let's talk about operating them.

Step number one is to make sure there's a clear path between you and an exit. No matter how small and insignificant a fire may seem, don't try to fight it if it's burning between you and your doorway to safety. If a fire spreads, spills, or falls between you and your escape route, forget about trying to extinguish it and leave immediately through a secondary exit. Flash point could occur momentarily, and you could be trapped.

Most home fire extinguishers have some sort of safety device that prevents accidental discharge if they are knocked over. Some have a tape or a piece of plastic that must be broken before the trigger is operational. Others have a plastic pin that you must pull out before the trigger can be depressed. (Don't wait until you're faced with a fire to find out how your extinguisher works. Read the instructions carefully before making a purchase, and periodically reread them to keep the procedure fresh in your mind.)

With the safety disconnected and your back to your escape route, you are ready to attack the fire. Fire extinguishers discharge with a mild "kick," so brace yourself. It's doubtful that anyone, even a child, would be knocked over by such a reaction—it isn't *that* strong—but when combined with the loud noise of the propellant gas being released, the whole thing may be startling. Plant your feet firmly and grip the extinguisher tightly to avoid inadvertently dropping it when it discharges.

Face the fire and aim the nozzle at the base of the fire. That's where the source of combustion is, not up in the cloud of smoke above it. Using a sweeping motion, spray back and forth across the fire's base, covering any sparks or spills that may be surrounding the main fire itself.

An extinguisher will not discharge indefinitely. Most home units contain only about ten to twenty seconds' worth of extinguishing material and propellant. That may not sound like a lot, but the chemicals in the extinguisher are very effective and can put out a rather sizable fire if you apply them properly. In order to be even more effective, it's best to discharge the extinguisher in several short blasts rather than one long one.

If you find you've misjudged both your ability and the capability of your extinguisher to put out the fire, don't try to find another extinguisher—there isn't time. Get out at once and call the fire department.

Assume you've been successful in putting out a small fire with your extinguisher. Your job's not over yet. Go through all the steps listed on page 54 to be sure it's out. And remember, call the fire department after extinguishing a fire (except for well-contained wastebasket or small cooking fires) to make sure it's out *for good*.

What to Look for When Buying a Fire Extinguisher

Manageable weight	Ease of operation
Convenient size	Pressure gauge
Safety mechanism	Bracket for hanging

Home fire extinguishers usually weigh five to ten pounds (about half of that is chemical, half is container). Most of

-2 Aim at base of fire and spray with a sweeping motion.

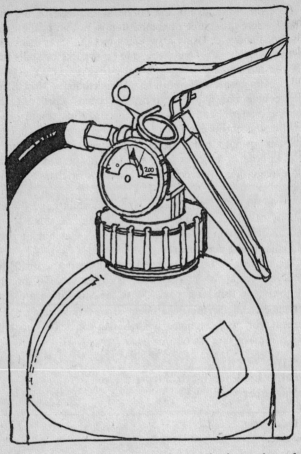

6-3 Extinguisher with a pin-type safety mechanism and a safety gauge.

them are cylindrical in shape and measure sixteen to eighteen inches high. The lighter-weight ones are about three inches in diameter; heavier units may be four to five inches in diameter. Make sure everyone in your household can handle the unit.

Another factor to consider is the safety device that keeps the extinguisher from firing if dropped. The kind that requires the operator to squeeze a handle tightly enough to shear off a plastic pin may be difficult for some people in your family to use. Choose a unit instead that is activated by simply pulling out a restraining pin or breaking a piece of tape. Don't try to protect your family with equipment that's tricky to operate.

Select only an extinguisher that has a pressure gauge. That way you don't have to wonder whether it will really work when you need it. The chemicals in an extinguisher are under pressure—that is, in addition to whatever chemical mixture the unit contains, there is a gas inside that will cause the chemical to shoot out when the trigger is depressed. It's possible for the propellant gas to leak out, making the extinguisher useless. A pressure gauge tells you at a glance whether there is sufficient gas in the can to make it work. If the pressure should ever appear low, have the unit recharged at once.

It's a good idea to buy an extinguisher that has a bracket for hanging on the wall. Use it.

Where to Put Extinguishers

Assure escape routes.
Allow unobstructed access.
Make them reachable by all.
Put them where they're needed.

If the need ever arises for you to grab a fire extinguisher, you won't have time to go searching for it. So hang it out where you can get to it quickly and easily. (Keeping it out in plain sight reminds you to check the pressure gauge periodically, too.)

A good place for an extinguisher is right next to an outside door. Every home is different, of course, and you have to decide upon the best place in yours. Just make sure you guarantee an exit—don't hang your extinguisher in a dead-end corner where you could become trapped going after it.

Never put an extinguisher over the kitchen stove where you might have to reach through flames to get to it. Instead, hang a kitchen extinguisher on the wall opposite the stove so that if a fire starts on the rangetop, you can stand back from the flames with your back to the door and still reach the extinguisher.

Young children can be taught to operate fire extinguishers. In fact, your life and theirs could someday depend on their knowledge of fire extinguishment. Don't make it hard for them—hang the extinguisher within their reach.

A fire extinguisher in or near the kitchen is a necessity. Other spots you might consider are near the bedrooms, the garage, and the workshop area. Since fires in each of these lo

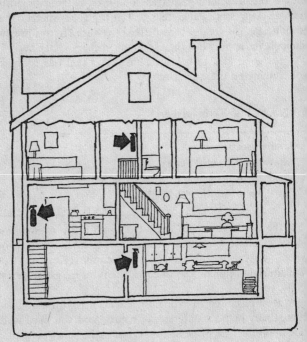

6-4 Location of extinguishers in a multilevel house.

ations could easily involve multiple classes of combustibles, choose extinguishers that can be used on A, B, or C fires. That way you won't waste valuable time deciding whether you have the right kind of equipment at hand and, worse yet, you won't risk making a bad situation even worse.

7. PREVENTION OF HOME FIRES

Which country in all the world would you guess has the highest number of deaths resulting from fire every year? China, with its enormous population? Mexico, with its crowded slums? Japan, with its wood and paper houses? No, it's the United States that holds the record—8,000 people killed in fires each year. On a per capita basis, our fire death rate is rivaled only by Canada, and it's almost twice the average of the rates for a number of European countries.

How can that be? In part, it's the price we pay for being the most industrialized and advanced country in the world. All our sophisticated fire-fighting equipment, extensive communications systems, and modern hospitals can't offset the fire hazards presented by our high scale of living and the carelessness and complacency it seems to engender. We take so much for granted, and disregard the most elementary and commonsense precautions. Fire regulations are not strictly enforced. Yet every electrical appliance is a potential source of fire; many chemical products we use daily are as dangerous as time bombs; air-conditioners and central heating create special problems; and even that home standby—television—is to blame for numerous fires.

Let's take a look at some of the common hazards likely to be found in the home. First, there is one danger so widespread and so lethal that it has to rank above all others—smoking.

Smoking Rules

When feeling drowsy, put your cigarette out.
(This is especially important after drinking.)

Don't smoke in bed.

Use large ashtrays with tight-fitting holders.

Before retiring, empty ashtrays, preferably into a can
   of water.

Don't smoke in a home workshop or anywhere you
   use flammable materials.

Of all the various causes of residential fires, nothing kills
or injures more people than the careless use of smoking materials. Most of these deaths result from inhalation of toxic
gases generated by smoldering cigarette fires in mattresses
and upholstered furniture. Many people die in their sleep, the
victims of their own careless smoking habits. Tragically, many
others also die as victims of those smokers' negligence.

Can these deaths be prevented? Of course they can. Fires
started by smokers are not natural disasters over which we
have no control. They are fires caused directly by people, and
they can be prevented by people.

Cigarette manufacturers, along with users, must share some

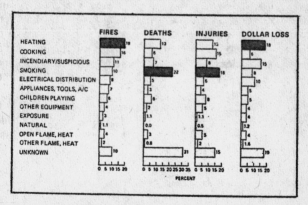

7-1 Causes of residential fire losses. (From *Highlights of Fire
in the United States,* 2nd ed. Federal Emergency Management
Agency/U.S. Fire Administration: NFIRS, 1978.)

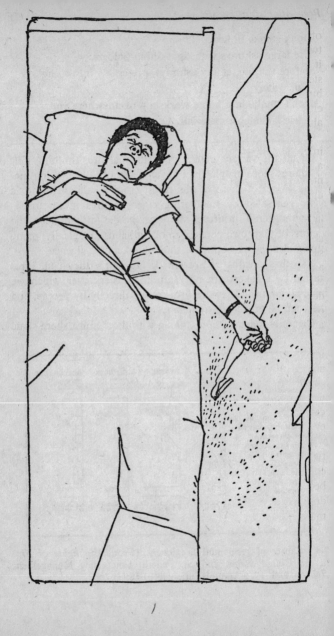

of the blame. To ensure that a cigarette will stay lighted between puffs, they mix an accelerant into the tobacco. Without it, a discarded or dropped cigarette would quickly die out instead of continuing to burn and possibly igniting other combustibles. Efforts so far to convince the tobacco industry that cigarettes should be self-extinguishing have been unsuccessful, although a few brands are available that go out quickly.

I recall one particular smoker's death that was staggering in its simplicity. We found the victim dead in bed without a mark on him—no burns, no evidence of smoke inhalation, no sign of his having awakened. Before going to bed, he had apparently dropped a cigarette down into the cushions of an upholstered chair next to the bed. The fire smoldered and burned for several hours and finally totally destroyed the chair. That's all, just the chair. The man's room was so tightly sealed that the fire had used up the entire supply of oxygen, and then, with no more oxygen to help it burn, had gone out as quietly as it had started. The man was asphyxiated, deprived of oxygen to breathe. When we opened the room, everything was normal except for the remains of the chair and the presence of one more fire victim.

In a large percentage of the fires caused by careless smoking habits, alcohol contributes to the tragedy. Even a small amount of alcohol can impair judgment or cause drowsiness —or both. Heavier drinking can, of course, lead to forgetfulness, poor coordination, and unsteadiness, any one of which can mean the end for smokers and those around them. However, you don't have to have consumed alcohol to doze off while holding a cigarette. We sleep in stages, alternating between light and deep sleep, with the first stage of deep sleep coming approximately forty-five minutes to an hour after we become drowsy—just about the amount of time needed for a dropped cigarette to smolder for a while, ignite a blanket and mattress, and begin emitting poisonous gases.

The person who insists upon smoking in bed owes it to

7-2 NEVER smoke in bed.

himself and to everyone else in his household to install a smoke detector over his bed. To do anything less is just plain foolish.

Cute little ashtrays may look pretty sitting on an end table, but they don't really serve the purpose. Ashtrays should be large enough to accommodate several cigarettes and should include a fairly tight-fitting cigarette holder. If a cigarette is wedged into a holder between puffs, it is more likely to extinguish itself if it is left unattended. Make it a practice to gather up all the ashtrays before going to bed and pour their contents into a can of water. This is especially important after smoking guests have been in your home—doubly important if you've all been drinking.

The living-room fireplace is no place for discarded cigarettes. They could smolder there undetected, igniting the unburned remains of an earlier fire and, if the flue is not open, spread suffocating fumes throughout the house.

Smoking is dangerous at any time in a home workshop, where there is apt to be a plentiful supply of combustibles, particularly materials such as paints, glues, and solvents. But wherever you use such substances, refrain from smoking.

Although they don't contain an accelerant as cigarettes do, cigars and pipe tobacco can be a problem. A carelessly knocked-out pipe may scatter ashes that can hide and build into a larger fire.

Other Household Hazards

Furnaces and heaters	Candles and paper decorations
Cooking stoves and ovens	Matches and lighters
Electrical appliances	Trash and other burnables
Electrical wiring	
Flammable liquids	
Aerosol cans	Power lawn mowers
Television sets	Burglar bars, bolts, and chains
Fireplaces	

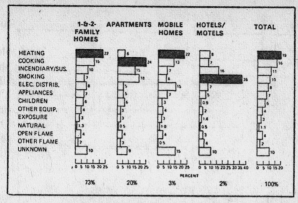

7-3 Causes of fires by type of residence. (From *Highlights of Fire in the United States,* 2nd ed. Federal Emergency Management Agency/U.S. Fire Administration: NFIRS, 1978.)

FAULTY FURNACES AND HEATERS and careless use of them account for the largest percentage of residential fires every year. The rate is highest in the southern states, where many homes don't have central heating units and instead use stoves and space heaters. Often fires are caused in those circumstances by pushing beds too close to the heater. But the colder, northern states also have an abundance of heating-related fires.

In recent years there has been a return to the use of wood stoves as an auxiliary, or sometimes primary, source of home heating. This has led to some real problems (22,000 fires and 800 deaths and injuries in 1980 alone) because people new to their use may not take the same precautions their ancestors did. For example, stoves should be set on fire-resistant bases, flues should be periodically cleaned, and only certain kinds of wood should be used. County agricultural extension agents are sources of accurate and helpful information.

COOKING STOVES AND OVENS are the source of many fires, particularly those due to carelessness. A neighbor

tops in, the telephone rings, or your favorite television show reaches its most exciting moment, and you forget all about dinner blazing away in the kitchen. Food burns—it *really* burns! Grease will burst into flame when heated to its ignition point; the bottoms will melt right out of aluminum pans, spilling the contents over the top of the range; a single pan of vegetables will give off enough choking smoke to fill a kitchen. If you think there is any chance of forgetting something you've left on the stove, set a timer to go off as a reminder.

Take a careful look at and around your kitchen stove. Are there potholders hanging over it? Is there an overhead exhaust fan covered with grease and dust? Do you have curtains hanging nearby, or is there a paper-covered shelf next to a burner? Any one of these things is a fire risk.

Outdoor cooking can be fun, and it can be dangerous. First thing: Keep children away from the fire. Charcoal burns with an intense and lasting heat. The coals may not appear to be burning at all but may still be very, very hot. Anything combustible (toys, sticks, paper, dolls) dropped onto them can suddenly burst into flame. Charcoal starter is intended to be just that, a starter. Once you have soaked the coals with it and they have begun to burn a little, don't try to speed things up by spraying more fluid on them. Charcoal starter is kerosene. It can vaporize over the small glow and explode, causing the flame to travel back along the stream to the can you're holding.

Be careful where you set a barbecue grill. Look all around and *over* the grill to make sure nothing combustible is near enough to catch fire. Don't try to cook on the back stairs or balcony of an apartment building. If the grill were tipped over, it could not only set fire to porches, stairs, and balconies, but also eliminate a means of escape for anyone trapped in the building.

4 Use a barbecue grill well away from combustible material or overhangs such as an arbor.

MOST ELECTRIC APPLIANCES sold today carry a
Underwriters Laboratories-approved label. That means th
appliance has been tested and has been found to be safe *i
used as directed*. It's that last phrase that's important. Som
appliances have warning tags that say they shouldn't be im
mersed in water; some warn that the case surrounding a
motor may become very hot; some state that they should b
plugged in alone and not used with extension cords or multi
ple outlets. Follow the directions!

HOME ELECTRICAL WIRING is generally safe if in
stalled by a professional electrician according to moder
building codes and practices and used properly and prudently
That's a big "if," and when it's abused, tragedy is often th
result. It may look like a bargain when your friend Joe offer
to rewire your basement so that you can turn it into a recrea
tion room. It won't seem like such a good idea when a spar
behind the paneling sends the whole place up in flames.

Overload is the most common cause of electrical fires. B
plugging three or four lamps and appliances (particularl
those that are heat producers such as irons, toasters, waffl
irons, heaters, and hair dryers) into the same outlet, you'r
just asking for trouble. Maybe your home is indeed under
wired and has too few outlets—that applies to many olde
homes built before "luxury" items became "necessities." Yo
can't afford not to have it rewired. If that isn't possible (a
in an apartment building), then at least learn to use only on
appliance at a time. Don't use multiple-outlet "octopuses"
and extension cords to try to beat the system. It's only a mat
ter of time before you create an overload that may lead t
fire.

FLAMMABLE LIQUIDS such as solvents and cleaner
must be in tightly sealed metal containers, away from hea
Terrible tragedies have resulted from people using the

7-5 An overloaded electrical outlet.

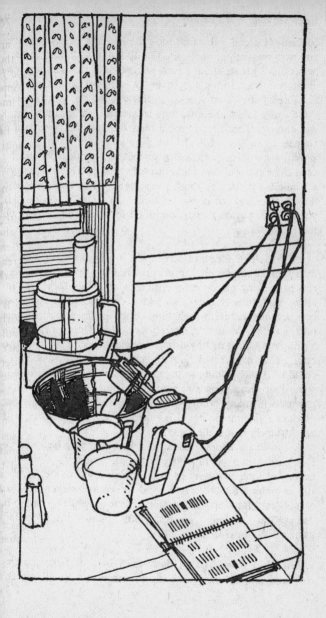

foolishly. I recall a fire that took the life of a pretty teen-age girl who was helping with the family laundry. Before putting her father's grease-stained work clothes into the washing machine, she tried to get the spots out with gasoline. Gasoline! She knew better than to use gasoline around an open fire, but there was no flame nearby, only an electric water heater. That was enough. The laundry room blew up, and she never had a chance.

When the label on a can says that the product should be used only in a well-ventilated area, believe it! Don't just open a window a crack and think you're safe. Make sure there is a flow of fresh air *through* the area where you're working. If you're cold, put on a coat, but don't let dangerous vapors fill the room.

AEROSOL CANS are one of the wonders of our age. So many things can be packaged in them, everything from toiletries to cooking oil to bathroom cleaners. While they may increase convenience, they also add danger. Some of the products are flammable in their liquid state—spray them out as a mist, and they become extremely explosive. Hair spray is such a product. A mist of tiny droplets released near a lighted cigarette or an overheated hair dryer could suddenly become a cloud of flame. Adding to the danger is the propellant used in some hair-spray cans—propane gas. That's the same kind of gas used in many bottled-gas cooking stoves. So in this one product a flammable liquid is released into the air along with an extremely flammable gas.

Aerosol cans should never be stored near any heat source. Don't set a can on or above a radiator; next to a furnace, heater, or kitchen range; or even in a spot where the summer sun can shine directly on it. And, of course, never drop an empty can into an incinerator, because while it may be "empty" of the liquid product, it still contains the propellant gas and could explode.

TELEVISION SETS. A number of fires start in *television sets,* and either turning off the power source (wall switch)

7-6 Hair spray, a hair dryer, and a lighted cigarette—a dangerous combination.

or unplugging them at night is an easy and sensible safety precaution. If you have an outlet that is controlled by a wall switch, that's a good place to plug in your television set. Then every night when you go to bed, make it a habit to turn off the wall switch as well as the set. This is particularly important if your television set is the "instant on" variety. "Instant on" really means "always on." The set is actually on at all times; the power is just turned down when you're not watching it.

Two more things to remember about television sets: Don't put a television flat against a wall or in a close-fitting enclosure such as a bookcase (it generates heat and needs lots of air around it), and never throw water on a burning set, *even if it is not plugged in*. (Television sets have the capacity to hold electricity even after being disconnected and can explode.)

FIREPLACES are valued as much for the romantic glow they give off as for the warmth they provide. Don't let yours become a hazard. Set a tight-fitting wire-mesh or glass screen over the opening and place a fire-resistant rug on the floor in front of it. Don't use the fireplace as a trash receptacle and don't throw lighted cigarettes and matches into it if there's no fire—you could accidentally start one when you don't want to. If you have the kind of fireplace that has a chute in it for collecting ashes, make sure all the fire material is completely out before dumping it down the chute. There are several other things to keep in mind with a fireplace: Don't leave a fire unattended, don't use a fireplace for cooking, don't use artificial logs, don't burn the wrong material, don't allow children to play nearby, and don't hang decorations over it.

Never go away and leave a fire burning in a fireplace. Make sure it's completely out before you go to bed. Fires have a way of coming back to life when you don't expect them to.

Setting a small charcoal grill in the fireplace to broil a couple of steaks could result in a much more expensive meal than you anticipated. A tipped-over grill could spill grease and red-hot coals onto a flammable carpet. Also, charcoal gives off carbon monoxide. Don't risk it! (The same caution applies to those little tabletop charcoal burners for heating hors d'oeuvres—they're just too risky.)

You should avoid using paper-wrapped artificial logs, for two reasons. They are made by soaking wood chips in paraffin (extremely combustible) and then shaping the mixture into log forms. They produce considerably more heat than real logs and, if stacked up in a fireplace or covered with real logs, can generate enough heat to explode, blowing the fire out into the room. Also the paraffin contained in the log produces a greasy smoke that can coat the inside of your chimney and become a fire hazard. (*Never* use artificial logs in a closed wood stove!)

Don't use your fireplace to get rid of trash. I recall one man who thought he'd found an easy way to dispose of his Christmas tree without spreading needles throughout the house. He put the whole tree in the fireplace, small end first, and threw in a match. The dried-out branches caught with such speed and intense heat that the fire literally exploded out into his living room.

Christmas boughs and paper chains may look pretty draped across a mantel but can be ignited by only one small popping spark from the fireplace below. Many screens cover only the front of the fireplace, not the top. Even without a spark, the heat alone could be sufficient to burn nearby decorations, furniture, or curtains. Make sure nothing flammable is near enough to the fire to catch.

Keep an eye on the kids when you have a fire burning. Children are fascinated by the dancing flames and may try to poke at them. Don't leave a child alone for even a minute with a fire. Clothing can ignite in an instant. Impress upon your entire family the danger and need for extreme caution around the fireplace.

CANDLES AND PAPER DECORATIONS make for a festive occasion. They also make for a potentially dangerous

one. If you must use paper decorations, don't have a lot of them—choose another material that doesn't burn. As for candles, don't put them in little wood-block or Styrofoam candleholders that tip over easily—anchor them firmly into heavy candlesticks or set them deep down into sand-filled bowls. Never set candles inside bookcases or under anything that will burn.

MATCHES AND LIGHTERS have their own attraction for children. Just as children are intrigued by fire in a fireplace, they are fascinated by small flames and are eager to get their hands on matches and lighters in order to make their own. Telling them repeatedly not to play with matches is not enough. Think about the consequences of a child guiltily hiding in a closet and striking a match. You're better off—and so is your child—if you demonstrate the proper use and handling of matches and talk to him or her openly about the danger of carelessness and playing with fire. Actually taking children to view the remains of a house fire can be a valuable object lesson.

TRASH AND OTHER BURNABLES can accumulate in your home in a very short time. The saying that "fire doesn't burn in clean surroundings" may be an oversimplification, but to a large extent it's true. By eliminating fuel sources and opportunities for fire to occur and grow, you'll go a long way toward avoiding disaster. You may want to keep some rags for cleaning or some scrap pieces of lumber for repairs. There's nothing wrong with that as long as you don't allow oily, greasy rags to lie around or sawdust to build up. Keep only dry, clean rags and neatly stacked lumber, as dust-free as possible.

POWER LAWN MOWERS must be operated with caution, not only because of their sharp blades but also because

7-7 Teach children the dangers of playing with matches.

of the gasoline they contain. When your lawn mower runs out of gas, don't try to refill it until it has had a few minutes to cool off. Gasoline spilled onto a hot engine will instantly vaporize and can burst into flame. It could ignite the gas can you're holding. Naturally you should not refill the mower inside a garage or shed. Take it outside where you're safely away from other combustibles.

BURGLAR BARS, BOLTS, AND CHAINS are a potential household danger of a special category—they are there for safety reasons in the first place. I recall an elderly physician whose home was in a wealthy but remote part of town. He was surprised one night by a gang that ransacked his house and left him badly beaten. Determined to protect himself in the future, he installed bars on all his windows and wired the whole place with burglar alarms. What he forgot was the enemy that came from within his home—fire. He died in his own fortress, unable to escape through a window when flames came rushing into his bedroom.

People are afraid. They install complicated locks and bars on their doors and windows to prevent thieves and vandals from getting to them. Windows in apartment buildings are barred so that small children won't fall out. But elderly folks and small children are likely to be the ones who die when a fire starts in the house and there's no way out.

Burglar bars are fine—*if* they can be released and opened from the inside, and *if* everyone in the household—including the children—knows how to work them. Key-operated dead bolts are good protection—*if* the key is hanging within reach of every family member, and *if* the family fire drill has included instructions and practice in unlocking them. Lock bars are wise to have on sliding glass patio doors—*if* they can be easily and quickly removed.

There's no point in protecting your family from one kind of danger by creating another. A house that you can't get out

7-8 Show children how to open burglar bars.

of is nothing more than a cage. You wouldn't lock yourself in with a man-eating tiger, would you?

As lengthy as this list of household hazards may appear to be, it's not complete. Look around your own home and you'll undoubtedly find some more. In fact, it's a good idea to make a periodic walk-through of your entire house with a critical eye. Perhaps a small table sitting in front of a patio door could be moved to one side, or a bag of knitting next to a fireplace could be stored elsewhere. What about the drapery cord next to the window fan—could the cord blow into the blades, jamming them and causing the fan to overheat, setting fire to the wall covering? Watch for electrical hazards (a lamp cord caught under a chair leg, or a curtain hanging over an overloaded outlet) and be especially alert for potential fires associated with flame and heat sources (potholders hanging near range burners, towels draped over a heater to dry).

Care and prevention are your best safeguards against fire. If there is any question in your mind about the safety of a situation, take action immediately to correct it. Smokey the Bear's warning applies at home as well as in the forest: "Only you can prevent fires!"

8. DRILL! DRILL! DRILL!

Somebody has to take charge.

Somebody has to decide that your family is going to be prepared should your home catch fire. Somebody has to plan ahead and organize escape routes and procedures for each member of the family—and do it so well that if the time should come to use his training in an emergency, everyone will react automatically to save himself. That somebody is probably you. You can't afford to wait until somebody else gets around to formulating a plan. Your home could catch fire next year, next week, or this very day. A plan delayed is a plan unmade.

You can read all the books and magazine articles you want about how to get out of a burning building, but until you actually sit down and figure out how to escape from *your* house or *your* mobile home or *your* apartment, none of that general information is going to do you a bit of good. Your plan has to be suited to your particular home and to all the people in it.

Imagine yourself out on a boat during a storm. Suddenly the boat begins to take on water, and you know it's going to sink. Now, you've never been a swimmer, but you've read a lot about swimming. You know that you have to kick your feet, stroke with your arms, and keep your head out of the water in order to breathe. That's not so hard—you've seen people swimming lots of times, and it always looked pretty easy.

Then you find yourself in the water, gasping for air. Where is the shore you were so close to a few minutes ago? The storm has made everything dark, and the waves are so high that all you can see is water, water, and more water. You hear the shouts of your family, but you can't find them. Your lungs are aching from holding your breath. You try to remember all those stroke-kick-breathe directions, but you keep getting mixed up, and the waves are constantly pounding against you. You feel the water closing over you. Maybe you have time to think, What a stupid way to go!

Escaping from a burning building is very much like this situation. Smoke may be so thick and so black that you have no idea in which direction you're going. Familiar objects suddenly feel strange when you reach out in panic. You think about helping your loved ones, but you can't even seem to save yourself. You can't catch a breath of air, and you can't quite get your bearings. Which way is the door? If you could just *see* or get a little fresh air! You know you have to do *something,* but you're disoriented and can't remember how to make your arms and legs work the way you want them to. You feel yourself going down.

With proper training and a few precautions, this scenario doesn't have to happen this way. You can learn to act rationally and calmly when fire strikes. But you have to work at it, and you have to be willing to practice ahead of time. All the reading in the world won't help you if you haven't made your plan and run through it regularly to familiarize yourself with each step of your escape procedure.

For an Effective Home-Evacuation Plan

Take charge! Don't wait for somebody else to get around to it.

Draw a complete floor plan indicating all exits (doors, windows, halls, stairs), including all interior doors, closets, storage areas, and blocked exits.

Color-code primary exits green; secondary exits yellow; dead ends red.

Plan each person's first, second, and perhaps third escape routes.

Assemble the entire family, even toddlers and the elderly.

Emphasize seriousness and tolerate no playing around.

Discuss your plans and ask for each one's participation.

Talk about how all of you will escape.

Avoid overloading young children with too much detail; concentrate on escape routes from their bedrooms.

Assign an outside meeting place.

Caution against reentering the building for any reason.

Describe the procedure for reporting a fire.

Repeat, repeat, repeat everyone's first concern: Get out and away from the building.

A good way to plan your home escape drill is to have a diagram of each floor. If you aren't very proficient at sketching, pick up a few sheets of graph paper.

Going from room to room, draw a floor plan that includes every window, door, hallway, and opening between rooms. Don't leave out closets, pass-throughs, pantries, and storage rooms or storage areas. Indicate the locations of extinguishers; furnaces, stoves, heaters, and water heaters; air-conditioners; fire escapes; skylights; barred windows or doors; and all stairways. Some older homes may have dumbwaiters, or windows in unusual places such as in closets or under staircases—put them in the plan.

If you live in any kind of multiple-family building or in an apartment connected to commercial or industrial space, add all the common-area entryways, corridors, elevators (passenger and freight), stairways, windows, and fire escapes. There may be additional exits that you could reach from your apartment by going through other unlocked areas on your floor. Think about every single way out of the building that you can and include it on your diagram. When you draw a common corridor, don't forget to include storage-room doors.

There may be some windows or glazed doors in your house or building that are barred or permanently nailed or painted

8-1 Make a floor plan.

shut. Obviously they can't be considered primary exits, but
they should be drawn on your plan. Doors and windows leading onto balconies, ledges, and rooftops are not first-choice
exits unless they have fire escapes attached, but they could
still be pathways away from immediate danger. Put them in
your plan. It would be helpful to color-code your drawing for
emphasis. For instance, you might indicate primary exits in
green, secondary exits in yellow, and dead ends in red.

FAMILY
FIRE
PLAN

8-2 Give a copy of your fire evacuation plan to the baby-sitter.

Once you have diagrams drawn of each floor of your home, including the basement, you might want to make several photocopies of the plan, especially if you have frequent overnight visitors or baby-sitters. Hung on the wall next to a guest-room door, a copy could save the life of a visitor or, given to a sitter, could make the difference between whether your children live or die while you are away from home. (Be sure to stress to a sitter that your children are aware of the evacuation plan and have practiced it with you. Contradictory instructions by a sitter during an emergency will lead to confusion and delay.

After drawing your floor plan, assemble the entire family for a discussion and planning session. Everyone should participate, even toddlers and the elderly. You may be surprised at some of the misconceptions you hear ("We live in a single-story house, so we're safe"; "We have all electric appliances, an electric furnace, and none of us smokes, so we're safe"). On the other hand, you may find that your children can share with you some very useful information they have already picked up from a fire-department representative who visited their school.

Whatever ideas and suggestions come up, take the time to talk about them. A calm, well-reasoned, open discussion can help bring to light potential evacuation problems or unsuspected hazards. It also gives you the opportunity to stress the danger you all would face in the middle of the night in a smoke-filled house. Naturally, you would not want to be so dramatic that your children would have nightmares.

A careful explanation of exactly why you should be prepared to take certain actions and avoid others when an emergency arises is necessary at this time. It's not enough simply to tell everyone to stay close to the floor in a smoke-filled area. Children especially may have a difficult time imagining what a whole house filled with smoke is really like. Draw some pictures to show them how smoke and fumes tend to rise and gather at the ceiling while the air closer to the floor remains clear. Explain why it's important to crawl rather than run to an exit, and why it's even better to keep a damp cloth

FAMILY FIRE PLAN

3 Go over the plan with the whole family.

over your mouth and nose in order to filter out as much smoke as possible while crawling.

Your first training session should be a complete lesson in the way a fire burns and spreads, what hazards and fuel supplies exist in and around your home, and when and where are the most likely times and places for a fire to start. Point out that fire prevention is of critical importance, but that, no matter how careful everyone is, there is still the chance of fire striking your household. Inform your family that you are going to conduct periodic evacuation drills, some announced and some without warning, so that everyone can become accustomed to doing exactly what is expected of him or her when an actual emergency arises.

You may have to emphasize repeatedly the serious nature of your family training and drilling. *Drills are not games* and should not be treated as such. Never allow your drills to become opportunities for joking and horsing around. Maintain order and show by your example that you consider evacuation practice to be important business.

After explaining your home's complete floor plan and discussing how fires can start and the ways they spread, take the whole family on a walk-through of your entire home. With the floor plan in hand, go carefully and slowly through each room pointing out every single exit: windows, doors, stairways, corridors.

Talk about the ways in which you might become aware that a fire has started and how each of you would react. What would be the best action to take if a curl of smoke were seen drifting upward from a wastebasket? What would you do if you were awakened by the buzzing of a smoke detector and found the room totally filled with smoke? How would you react to flames and smoke pouring out of the broiler while dinner was being prepared?

As I pointed out in chapter 5, there are times to fight, and there are times for flight. Each adult family member who is physically able to do so should know how to smother a small fire and how to use the household extinguishers. Emphasize the need for careful judgment in determining whether to t to put out a fire or whether to escape. Children sometimes get the idea that even though they are completely innocent, they might be blamed for a fire getting started when no one else is around. Fearing punishment, they may attempt to extinguish a blaze that is much too big for them to handle. If they're to survive, you have to emphasize that they should escape—that they are far more important than anything else that might be destroyed.

You've laid the groundwork; now it's time to run through a complete drill. Since most home fires start at night, have everyone go to his bedroom and wait for your signal.

When everyone is in place, call out "Fire!" At that signal, each person should drop to the floor and begin crawling to the nearest exit. On reaching the exit, everyone should leave the building and meet at a prearranged spot outside and well away from the building.

8-4 Practice crawling to the nearest exit.

An exercise like this will probably take several minutes. That's okay. The first time you want only to show the importance of keeping low and using the most convenient exit.

Now try it again, but this time use a wristwatch and see how fast you can all go through the same procedure. When you have finished the exercise, talk about what you have just done and discuss any problems that any of you might have had. Can you find any drawbacks to getting out this way? What if one of the exits were blocked by flames? What would you do if the weather were cold and you were in pajamas when the smoke detector went off? What if the doors leading into the hallway felt warm when some of you tried to leave your bedrooms?

Set up two or three hypothetical situations in which various exits are blocked and unusable. Using the lighted- or snuffed-candle technique, begin the drill by setting off your smoke detector. Try to come up with several "worst case" scenarios— turn off all the power, for example—and use them in drills. By all means, you and your family should get used to having periodic drills. Every six weeks or so, you should run through the evacuation exercises simulating various emergency conditions.

I can't emphasize too strongly that you have to go all the way in conducting home fire drills. If the best way out of your children's bedrooms is a drop from their first-floor windows, then teach them how to do it—show them how to sit on the edge, then hang from the sill, and finally how to drop to the ground. Naturally, you should catch them if the drop is more than a very few feet or if they are very young. Otherwise, let them do it themselves.

If you have window escape ladders, use them! Each member of your family should practice climbing down— barefoot, which is probably what you'll be if a fire starts in the middle of the night. The best escape ladders have spacers that hold them out from the wall so that you can get your

8-5 Small children can be dropped from a window and caught.

8-6 Hang from the sill before dropping to the ground.

8-7 Escape ladders should have spacers to keep the rungs away from the wall. Without these "standoffs," the descent is painful (top figure).

feet onto the rungs, and the rungs should be about an inch in diameter so you can rest your full weight on them when you're shoeless. (Smaller diameter rungs make for such a painful descent, you might lose your footing and fall, in which case you'd probably sustain greater injury than you would if you had no ladder at all.)

Above all, don't kid yourself by thinking that you can always use unfamiliar escape routes if and when you have to. Unless you have actually practiced the use of unconventional exits, you may be too confused and unsure of yourself to be able to remember what to do when the actual need arises. Only by going through the entire procedure are you going to be familiar with each step.

Become fire-conscious. It isn't easy to do if you've never been faced with a fire in your home. You may think you're already doing everything right, or that nothing bad could happen to you. Not true. Maybe you've been lucky so far, but that luck could fail you tomorrow. You have to train yourself and your family to be on the lookout for fire danger and to evacuate quickly and safely should that danger become a reality.

8-8 A deluxe model escape ladder with well-braced standoffs.

9. HOTELS AND HIGH-RISES

At one time when anyone spoke of skyscrapers, you assumed he was talking about New York or Chicago. Nowadays even medium- to small-size communities all over the country have high-rise hotels, apartments, and office buildings. This new vertical way of living has brought with it the need for new fire-safety procedures.

Hotel/Motel Fire-Safety Precautions

Ask for a room on a lower floor.

Inquire about fire safety.

Locate two usable exits near your room.

Locate the nearest alarm box.

Familiarize yourself with your room or suite.

Put the key in the lock or on a bedside table.

If fire starts in your room, get out immediately.

If you suspect fire, smell smoke, or see flames, report suspicions at once.

Pick up your key.

Check the door for heat.

If the corridor is clear, get out at once.

Avoid elevators; use stairs only.

If the corridor is filled with smoke, stay in your room.

If you are unable to leave, seal the door, call or signal for help.

Fill the bathtub, turn on the exhaust fan.

Don't jump—wait for rescue.

Every hour of every day of the year there is a hotel fire somewhere in the United States. Every hour of every day of the year over $10,000 goes up in smoke in hotel fires alone. Think about that—a couple of dozen separate fires and about a quarter-million dollars every single day of the year!

And the lives! Just three fires—the Stouffer's Inn of Westchester, north of New York City, and the MGM Grand Hotel and Hilton Hotel fires in Las Vegas—claimed over a hundred lives. Besides those who die, many others are injured to such an extent that it may take months or even years for them to recover.

Ironically, in recent years the worst hotel fires, in terms of deaths and injuries, haven't occurred in seedy old flophouses. They've been in new, modern buildings. You can see why if you look at the ways in which new hotels and motels are furnished. Lots of plastics and synthetics are used now: plastic furniture and wall coverings, synthetic carpeting and upholstery. These new materials burn at rates of up to a hundred times faster than natural fuels and produce twice the heat energy.

The heat, smoke, and toxic gases released in a "modern" hotel fire may be carried throughout the building by heating and air-conditioning ducts, false ceilings, and elevator shafts. Since many more people are overcome by smoke than by flames once a fire starts (most of the eighty-four people who died in the MGM disaster were actually *nineteen stories* away from the fire), you need to know how to protect yourself from this silent killer.

I could point out the value of requiring every hotel and motel in the country to install sprinkler systems in every corridor and smoke detectors in every room. But we live in the real world, and we all know those changes aren't going to come about overnight. Does that mean there's nothing you can do to protect yourself during a hotel stay? Not at all. There are some very simple precautions you can—you must—take the next time you check into a hotel. Summarized on page 104, they make a long list. But from this list it's not at all difficult to develop a few simple lifesaving habits.

First, don't assume that serious fires happen only in high-rise hotels. There have been many deadly fires in one- and two-story hotels and motels. Stouffer's Inn, for example, is only three stories high, but twenty-six people died in that fire. You can improve your chances of surviving a hotel fire, however, by requesting a room on a low floor. Why? The most modern fire ladders will reach only about one hundred feet, the equivalent of six or seven floors. If your room is on a side of the building approachable by a fire department's equipment, you may be more easily rescued. A room on the first or second floor would be even better because you might then be able to escape through a window and drop to the ground outside.

When you check in at the front desk, ask the room clerk about fire safety in the building. Are there smoke detectors in the corridors? (One in each room is even better.) Is there an intercom system to warn guests of danger? Does the hotel have a sprinkler system and how extensive is it (kitchen only, or is it throughout the building)? Don't accept a room at the end of a long corridor far from an enclosed stairway even if it's on the second or third floor. Why decrease your odds of getting out?

Don't feel that you're being pushy by asking such questions. Look back at the first two sentences of this chapter. You can't afford *not* to ask about safety. Hotel managers are much more likely to remedy bad situations if they think you're concerned and that their business depends upon it.

Once you get to your room, locate two *usable* exits. Elevators don't count—look for enclosed stairways that lead to the ground floor or to the outside. You might also consider a window that opens onto a lower roof from which you could drop to the ground. It isn't enough that only you know the exit locations—make sure everyone in your party also knows. Having found the nearest stairway, open the door and look

9-1 When you check into a hotel, count the doors from your room to the nearest fire exit.

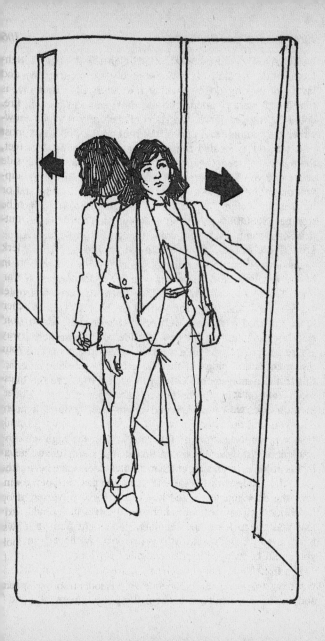

inside. Are the stairs unobstructed? Can the door be opened from inside the stairway? (Some hotels, for security reasons, have self-locking doors opening into stairwells. That means that if you were to attempt to use those stairs as an exit, you could be trapped inside the stairwell before you could reach either the ground-level exit or the roof of the building.)

After you've located the exits, count how many doorways there are between them and your room. If a fire were to strike and you had to crawl down the hall looking for an exit, you might not be able to see an exit sign hidden in the smoke up near the ceiling. When you're counting doorways, include storage and utility rooms so that you don't risk entering a dead-end room that you couldn't get out of again. Look around also in the corridor for fire-alarm boxes.

Inside your room, familiarize yourself with the placement of the furniture and be certain you have clear access to the door. Check the windows—do they open? Is there a fire escape or any other way for you to get out? If your window looks out into a window well or ventilator shaft, don't consider that an exit, because once outside your room there is nowhere for you to go. Take a quick look at the lamp and television cords and plugs. If they have broken insulation, report this to the management. When you go to bed, leave the bathroom light on with the bathroom door open a crack. You're in unfamiliar surroundings and might need a little light to help you find the door quickly.

A very important point: Put your key on the nightstand or someplace else where you can find it in a hurry if awakened in the middle of the night. Some hotel doors can be locked from the inside with the key. If that's the case in your room lock the door and leave the key in the lock. However, most hotel doors close and lock automatically. If a fire should start and you try to leave the building, you might find your exit blocked and have to return to your room. Without your key you could be trapped in a smoke-filled corridor.

If a fire should start in your room, don't try to fight it and don't try to save your belongings. You're in a strange place and have very inadequate fire-fighting equipment at hand—

probably a pitcher of water at best. Don't waste valuable seconds trying to put out a fire, even a small one. Leave at once, *close the door,* and either pull the nearest alarm or notify the desk immediately. Not only your own but a lot of other lives depend upon your reporting a fire as soon as possible.

Imagine waking up at three o'clock in the morning in an unfamiliar hotel room and smelling smoke. The fire isn't in your room, but there's smoke somewhere. Would you know what to do? Let's take it a step at a time.

Call the fire department. If your room has a direct-dial telephone, call the fire department and report your suspicions yourself. If all calls go through a switchboard, demand that the operator notify the fire department immediately, not just send someone up to investigate. Forget your belongings; there isn't time to gather them together. *Take your key!* Feel the door. If it's cool, brace yourself against it and open it a crack. Be prepared to slam the door hard if you see flame or heavy smoke in the hall. Cautiously look out into the hall. If it's clear or if there's just a little smoke at the ceiling level, stay very low and make your way to the nearest exit stairs. Close both your room door and the stairway door behind you. *Don't take an elevator!* Go down the stairs (hold on to the rail) to the ground floor and leave the building. If you encounter smoke as you go down, either return to your room or go on up to the roof and wait for rescue.

When you open your room door, you might see heavy smoke or even flame in the corridor. Don't try to go through it. Instead, close the door at once and stuff wet towels or torn sheets in the cracks under it. Do the same thing with any air vents that could direct poisonous gases into your room. If the telephone is working, call the desk and report your location.

Fill the bathtub with water and turn on the bathroom exhaust fan. The fan will help draw off toxic fumes at the ceiling level, and you can use an ice bucket to bail water from the tub to throw on the door when it gets hot. Keep the door as wet as possible. Most hotel doors are heavy and can prevent fire from getting to you for quite a long time; keeping the door wet is added insurance.

9-2 At night, leave bathroom light on and put key on night-stand.

Tie a damp towel over your face and stay as close to the floor as you can. If your room has double-hung windows, open the top a few inches to allow collected gases to escape and the bottom several inches to allow clear air to enter. If

-3 With ice bucket, throw water from tub onto door.

you can't open the window, just stay low. Breaking the window could allow heavy smoke or flames from the floor below you to enter your room. Wait as calmly as you are able to for help to arrive. Try not to give way to panic. Panicking can lead you to attempting something foolish. *Don't jump!* Rescuers will get to you. If you jump, your chances of surviving without serious injury are slim, especially if you're more than two or three stories up.

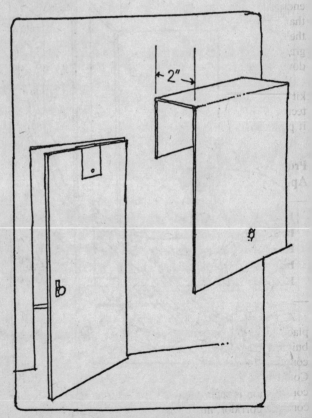

9-4 Hang your own smoke detector on bracket over hotel room door.

If you're a frequent traveler, you should seriously consider including a small fire-protection kit in your luggage. Nothing fancy—all you need is a smoke detector and a flashlight. You can either buy a travel smoke detector with a special door bracket or use an ordinary home smoke detector and make your own bracket. Cut a strip from a tin can and bend it into a square "C" that is two inches across the opening, large enough to fit over the top of your hotel room door and thin enough to allow the door to close. Punch a hole in the end that hangs down inside and hang your smoke detector from the bracket. Hang a flashlight from the doorknob so you can grab it quickly when leaving the room. It will light your way down a dark hall.

You might want to include a compressed-air horn in your kit to use as a signal. This is an extra, though. The smoke detector is the most useful item you can carry with you. Check it periodically to be sure it works, and carry a spare battery.

Protective Measures for High-Rise Apartments or Condos

Establish safety procedures and an evacuation plan.

Install smoke alarms and heat detectors.

Keep corridors and stairways clear at all times.

Insist upon safe practices from your neighbors.

Keep fire lanes open.

A well-constructed high-rise building is generally a safe place to live. My own home is on the forty-first floor of a building that rises another twenty-nine stories above me and contains a total of 900 apartments. Is it safe? I think so. Could it be evacuated if there were a fire? No, but then its construction is such that fire is unlikely to spread out into the common corridor, much less to another apartment, and I'd be safer staying in my own apartment if there were a fire in the building.

Every apartment's entry door is self-closing and seals tightly. It's made of steel and is set into a steel frame. The floors and ceilings are concrete, and firebreak walls separate every apartment on each floor. There are adequate enclosed stairways in each wing of the building and fire hoses on every floor.

Am I immune to fire? Absolutely not. Practically everything in my home could burn. I have appliances, air-conditioners, a water heater, a furnace, and a kitchen—every one of them a potential source of fire. I also have smoke detectors, heat detectors, and fire extinguishers because I don't believe in taking unnecessary chances. Neither should you.

It doesn't really matter that much whether you live on the fifteenth floor or the fiftieth floor. There are steps you can take to reduce the possibility of fire in your home and increase your chances of escaping injury or loss if fire should occur. These are summarized in the preceding box.

Just about all the precautions and procedures described earlier for hotels also apply to high-rise apartment buildings. If your home is in a high-rise, however, there are several additional things you can do to protect yourself and your family because you have more control over your surroundings.

Whether you live in a rental apartment, a condominium, or a cooperative, there is probably a building manager. Ask him or her what the prevention and evacuation plan is for the building—and don't be surprised when you're told there is none. Also don't assume that your inquiry is going to inspire the manager to put a plan together. If he shows a total lack of interest, you're going to have to take charge if it's going to get done.

Try to enlist the help of one or two neighbors to work with you in evaluating your building's strong and weak points. Once you have worked out a preliminary plan, call the fire department and ask if they will send over a representative to discuss fire prevention and safety with all the residents. Set up

9-5 Don't barbecue on a high-rise balcony.

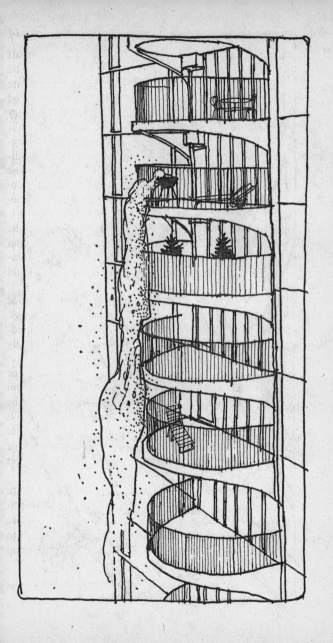

a definite time for a general meeting and begin publicizing it throughout the building about a month in advance. Contact as many people personally as you can to assure a good turn-out, and make certain the manager is present.

Work toward two goals in your meeting: Setting up an overall fire-safety plan for the building that will include scheduled drills; and establishing an ongoing safety committee of residents whose function will be to make certain everyone knows his responsibilities and is informed about the building's fire-safety policies. Make up posters showing all building exits and fire-extinguisher locations and place them on each floor and in the laundry room and storage areas. Distribute a news-letter giving fire-prevention tips, evacuation plans, and fire-reporting procedures. In this letter, impress upon the residents that the safety of everyone in the building is to a large extent dependent upon a cooperative effort—that is, each person's safety and that of all the members of his or her family depend upon the neighbors understanding the hazards of fire and knowing how to react in an emergency.

After you have started the ball rolling with your neighbors and the building manager, develop a set of plans for your own family. Refer back to the chapter on family fire drills; many of the items covered there apply to high-rises (and to other residences as well).

Install smoke and heat detectors in appropriate locations in your apartment, and encourage the building owners to place them in strategic spots throughout the building. It certainly doesn't hurt, either, for you to test common-area detectors periodically. It could be that they have dead batteries and no one has gotten around to replacing them. Report nonfunctioning smoke detectors to the building manager immediately.

A stairwell with a half-dozen plastic bags of trash in it is no longer a safe exit. Do your part to keep hallways and stairs free of any obstructions, and demand that the manager

9-6 Fire fighters with heavy equipment running up stairs of high-rise.

enforce similar conduct throughout the building. Never block any escape route or allow others to do so. Stairs, hallways, and exit doors are not storage areas and should never be used as such, even overnight.

Your building probably has "No Parking" signs posted around it to keep access open for fire-fighting equipment. Insist that those signs be obeyed. Precious minutes could be lost as firemen try to move cars blocking their approach to your burning building.

Because heating plants in large high-rise buildings are generally maintained professionally, heating fires are not nearly as common as fires originating from cooking. A barbecue grill on an open balcony is a high-rise hazard—if knocked or blown over, it can pour a stream of red-hot coals down on balconies and roofs below. If outdoor cooking is to be allowed for the residents of your building, it must be in a designated safe area. Never barbecue in an enclosed or covered area such as a hallway, stairway, storage area, basement, or garage.

Whether you're in a high-rise apartment building or hotel, remember that your safety depends upon how much you know. Learn all you can about your building, even if you're only staying overnight. If the building is your home, put together a fire-safety plan for you and your family—and practice it! Protect yourself with warning devices and easy-to-use fire extinguishers.

One last but very important point about high-rise fires. While you should never, never attempt to exit a burning building via an elevator, there are some elevators that can be used by firemen to get up to a fire. (The elevators have an override mechanism activated by a special key that enables the firemen to direct the car up to a floor below the fire floor.) If your building does not have such elevators, or if for some reason they are inoperable, then the firemen would have to climb the stairs to the fire. A fireman, even a young man in top physical condition, can move only so fast when climbing stairs and carrying fire-fighting equipment. On the average it will take him about one minute per floor. Consider this: If you live on the thirtieth floor, it could be a half hour before a rescuer could get up to you, so you must know how to do everything possible to survive for that period of time. Reading about it isn't enough; you have to practice. And you can't afford to put off working out your plan. You could need it tonight.

10. EXPENSIVE BARGAINS

People live in all kinds of places, from old brick row houses to suburban bungalows, and from sleekly modern town houses to renovated barns. A few of the homes are million-dollar mansions; most are much more modest. Living costs keep going up, and folks have to find ways of cutting corners. Since housing is often the largest item in a family's budget, lots of people start economizing there. It's possible, of course, to find adequate housing at a reasonable cost, but let's take a look at some "bargains" that can turn out to be far too costly in terms of safety.

Mobile Homes

Once called trailers or caravans, these houses-on-wheels have evolved into roomy and comfortable full-size homes. Too large to be pulled behind an automobile, most mobile homes nowadays are hauled by truck to mobile-home parks and set up permanently in one location. In years gone by, before any building codes applied to them, mobile-home fires were frequent and deadly. A few years ago, however, the industry adopted certain guidelines regarding construction and outfitting that have helped to cut fires and fatalities.

Modern mobile homes are generally ready to occupy when purchased. All you have to do is move in your clothing, dishes, and food, and you're in business. Basics such as draperies, carpets, wall coverings, and furniture are installed at the manufacturing plant. That's part of the fire-safety problem, though. As in modern hotels and motels, a lot of plastic and synthetic material is commonly used in the outfitting of mobile homes—materials that burn faster and at much higher temperatures than wood and plaster.

Before building codes, the temporary fire barriers found in conventional homes were often absent in mobile homes. Lightweight doors (often louvered or the accordion-folding type) and thin dividing walls covered on both sides with flammable paneling offered no protection at all. Electrical wiring was often not encased in metal conduit and was only lightly insulated. Windows were apt to be quite small, especially in the bedrooms where they would be most needed as exits during nighttime fires. If the windows were jalousies, they might as well have been barred since there was no way anyone could get through them.

Designed to provide a feeling of openness within a limited space, older mobile homes had less than ideal floor plans from a safety point of view. Because there was sometimes little or no barrier between the kitchen and the living area, a cooking fire could spread rapidly beyond the confines of the kitchen. Heating stoves and cooking stoves were often located adjacent to exterior doors. Any heating fire would immediately seal off that door as an escape route.

Bedrooms in some of these older mobile homes were located at the end of a hallway, away from an outside door. An arrangement like this, coupled with unusable escape windows, essentially doomed that bedroom's occupants during a fire. Owners of mobile homes with this kind of design should have escape doors installed that can be opened from inside the bedroom.

Look for the greatest number of safety features when you're shopping for a mobile home. Don't even consider one that doesn't have easy escape-route access from every room. Many mobile-home parks are located in rural or suburban areas that may not have the best fire protection—there may be no nearby fire hydrants, and the closest fire department may be understaffed, underequipped, and miles away. Home fire protection in the form of smoke detectors and *several* strategically placed fire extinguishers is absolutely essential in mobile homes.

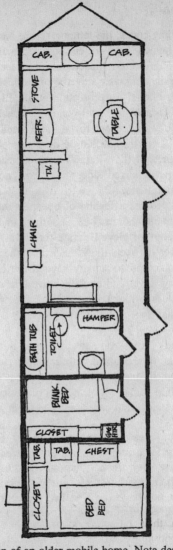

10-1 Floor plan of an older mobile home. Note dead-end bedroom and heater next to door.

Recreational Vehicles and Campers

Recreational vehicles, campers, and motor homes have the same potential problems as mobile homes—they may have floor plans that place cooking and heating stoves between the sleeping area and the lone door, and windows in these campers are almost never big enough for anyone to crawl through, even small children.

Here's a basic safety rule: If an occupied trailer is hitched to an automobile, the car should not be running. The exhaust fumes from the car could flow into the trailer, filling it with deadly carbon monoxide.

Don't build bonfires next to a camper; you could direct toxic gases into it. If your camper uses bottled butane or propane gas, a leaky valve could cause an explosion from a nearby fire.

Bonfires should be kept at a safe distance from tents as well. Throwing a log on a fire is fun—kids love to see the tower of sparks that shoot up. One of those sparks, caught by a breeze, is all it takes to set fire to a tent, turning a fun-filled weekend into a nightmare. Many tents are coated with waterproofing solutions that are highly flammable. So make it a rule never to take a lighted kerosene lantern or stove inside a tent—not if it's raining, not if it's cold, not for any reason. If such a stove or lantern tipped over or somehow happened to ignite the tent wall, you could find yourself trapped inside an envelope of fire. So when you go camping, take along one or two battery-operated lanterns. These oversize flashlights should provide adequate lighting for your needs.

Basement Apartments

Not too long ago three children lost their lives in a fire that raced through their basement apartment. Their bedrooms were at the opposite end of the basement from the only exit, and between them and the door was the kitchen, where an overloaded electrical circuit shorted out and started the fire.

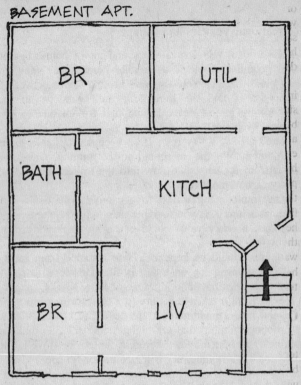

10-2 Floor plan of basement apartment with only one exit.

Neighbors and firemen tried frantically to get them out
through the windows but were unable to do so because of
burglar bars—put there to make the place secure.

There were at least four factors at work in that fire that
practically guaranteed tragedy: only one exit; bedrooms situ-
ated beyond the likely fire source; inadequate wiring; windows that were unusable for escape. Unfortunately, this kind
of basement apartment is more often the rule than the excep-
tion. I would guess that most basements used as residences
(many of them illegally so under city codes) were converted

or "renovated" by amateurs who did the work the cheapest way possible. Their intentions may have been good—to add some much-needed cash to the family income or to provide housing for relatives or friends—but they have ended up endangering the lives of anyone who lives in the apartment.

In such conversions, there is no overall plan. If the plumbing for laundry facilities is at the bottom of the basement stairs, that's where the kitchen goes, blocking the exit for the bedrooms that are back in a distant corner. Instead of hiring a licensed electrician, the amateur tries to do a "good enough" job himself. The tenant then moves in and installs his own multiple outlets and runs extension cords all over the place. It costs a lot of money to cut another doorway through the foundation, so the owner makes do with only a wooden flight of steps leading to an upstairs exit. For the same reason, he does nothing about the tiny windows that are six feet off the floor and don't open wide enough to crawl through anyway. Heat usually comes from inadequately vented gas space heaters or from electric ones that further tax the wiring system. The whole place is a potential deathtrap.

Over-the-Garage Apartments

Like basement apartments, apartments built over garages tend to be "handyman specials." They have the same drawbacks, plus more. Because they were probably not designed with safety in mind, there is apt to be only one exit, and the windows may be few in number and small in size. Aside from the obvious danger of carbon monoxide seeping into the apartment from an automobile running below, there are many opportunities for fire to break out, especially if the garage is used for any kind of hobby or workshop activity.

Suppose someone is refilling his lawn mower in the garage while you're peacefully watching television upstairs. He spills a little gasoline on the hot engine, igniting it. Startled, he drops the can and the spill spreads to a stack of cardboard boxes. Instantly the place is on fire. Add a parked car to the scene and you have the potential for an exploding gas tank

at could blow up the entire garage. Maybe you can't afford to pay much for an apartment, but you also can't afford to live in what amounts to a ticking time bomb.

Rooming Houses

Older, once-wealthy parts of cities with big, sprawling houses and neighborhoods around colleges and universities are usually filled with rooming houses. Tastes change, fortunes disappear, and owners begin taking in roomers in order to meet the costs of maintaining these fading mansions. Built to provide comfortable, even elegant, accommodations for a single family, these houses become the cramped—and potentially unsafe—living quarters of many individuals or families.

If fire breaks out, a beautifully carved open staircase may become nothing more than a chimney, spreading smoke throughout the house until it catches fire itself, eliminating an exit, and carrying fire to the door of every room. A dropped ceiling added to modernize some of the rooms serves as a perfect pathway for fire, permitting it to race through an entire floor in seconds. (Remember how fire travels: up, then across.)

Cooking and electric appliances represent special problems in crowded rooming houses. Although cooking may not be allowed in sleeping rooms, the rule could be and probably is broken in any rooming house. A neglected pan of soup or a teakettle could boil dry and start a fire, or a curtain or towel could come in contact with the red-hot coil of a hot plate. Television sets and irons are commonplace in such rooms—either could start a fast-spreading fire. Television sets can be especially hazardous in crowded rooms because people squeeze them into tight-fitting spots or drape a towel or piece of clothing over them, covering the air vents on the back and causing the set to overheat.

Rooming-house residents often are elderly people who may

be somewhat forgetful or have a little trouble caring for themselves. A lamp cord draped across a radiator or a towel drying on a space heater could start a fire that they might waste precious minutes trying to put out before calling for help. Older, partially infirm folks obviously are going to have greater difficulty getting out of a burning building as well.

The combination of old building, poor exits, crowded conditions, and perhaps physically handicapped residents makes for a dangerous situation. More often than not, very little is done to minimize it. If you or someone you love is a tenant in this kind of housing, the very least you can do is insist that steps be taken to increase the building's safety. There should be fire escapes—they may even be required by local housing laws—and they should never be blocked or locked. Smoke detectors should be in every hall, preferably in every room. There should be fire extinguishers available.

One of the greatest dangers associated with any of the apartments or mobile/motor homes discussed here is the limited means of escape if there is only one door. Every time you set foot into a single-exit home, you're betting that a fire won't start that will block your only escape route. And don't kid yourself into thinking there's another way through a second door if it's usually kept locked or is blocked by a piece of furniture. That's not an exit—it might just as well be a solid wall, because in an emergency you'll never be able to get it open.

Very simply stated, that low-cost but unsafe dwelling you've been considering for yourself or for someone you care a lot about may not be such a good deal after all. You could be looking at a bargain that would end up costing you heartbreaking memories for the rest of your life.

10-4 Crowded rooming houses may be unsafe, particularly for elderly residents.

11. FIRES IN CARS, BOATS, AND PUBLIC PLACES

So far, I've been talking about fire in houses, apartment buildings, and hotels. But, of course, fire can occur anywhere, anytime—in your car as you drive along a crowded expressway, during a pleasant afternoon's sail on the lake, while you're having dinner in your favorite restaurant. Learn to protect yourself wherever you are when fire strikes.

Vehicle Fires

Fires in automobiles are generally caused by a short in the electrical system. With more and more gadgets and extras in today's cars—power radio antennas, seats, door locks, sunroofs, and anything else you can think of to hook a motor to—the incidence of automobile fires has jumped alarmingly. Many fires are minor, but some can be extremely serious. For instance, a fire that knocks out the door-locking system and the window control could leave you trapped inside a burning car with no means of escape except to try to break a window with your shoe.

A fire extinguisher that's safe for use on Class B (gasoline) and Class C (electrical wiring) fires should be part of your automobile's basic equipment—as much so as a window scraper or a jack. The extinguisher has to be conveniently at hand, though, to be useful. If you have been carrying an extinguisher in your trunk, move it into the passenger compartment. In an emergency, you might not have time to get to the back of the car, fumble with your keys, and rummage around

11-1 A Class BC fire extinguisher being blasted through grille of smoking automobile.

under all the odds and ends that seem to accumulate in car trunks. And don't forget, the trunk area is often where gasoline tanks are located.

Install a bracket for the extinguisher under the dashboard, under the seat, or on the lower side wall of the front-seat section of the car. That way if a fire gets started, the extinguisher is handy to grab as you jump out of the car. (Don't carry it on the back window shelf—in an accident, it could fly loose and become an unguided missile.)

Just as in a burning building, getting out must be your first concern. At the first sign of fire, stop, turn off the ignition, either put the gear lever in "Park" or pull the hand brake, grab your extinguisher, and get out fast. That sounds like a complicated procedure, but running through it several times will help you develop it into a continuous flow of motions. No matter how small the fire, there is always the danger of involving the fuel line. At that point, the whole car can go up in a ball of fire.

Once you have mastered your technique for getting out of the car fast, explain the danger of car fires to your children. Stress the need to move fast and to get far away from the car. Perhaps you have an infant who normally rides in a special seat. That seat should have straps that unbuckle easily so that you can get the baby out fast. *Fast*—that's the key word in any automobile fire. The fire can spread fast and create what is literally an explosive situation. You have to be faster.

Once again, reading about and understanding the danger posed by a fire is not enough. You and your family have to practice. Only by running through a mock evacuation a few times can you be sure everyone really knows both the danger and what he or she is supposed to do.

How do you fight an automobile fire? From a distance. If the fire is under the hood—the most likely location—don't try to raise the hood. It might be stuck, or you could burn your-

11-2 A fire extinguisher should be basic equipment on boats of all kinds.

self. Instead, simply blast your extinguisher right through the grille. A Class BC extinguisher contains dry chemicals that will smother the fire inside without damage to the engine. And after the fire is extinguished, the powder can be blown off with a compressed-air hose. With the fire extinguished, call a garage for help. Don't try to drive the car again until the problem that caused the fire has been corrected.

Fires can occur in trucks, motorcycles, farm equipment, and other vehicles that have internal combustion or electric power. Whatever the kind of vehicle, the most important thing to remember is: Get out and away from the fire as fast as possible. Don't try to fight it until you're at a safe distance.

Boats

Boat fires are different from other fires in a very special way. You can't simply step away from a fire if you're out in the middle of a body of water.

In power boats, the same fire-fighting procedure applies as in an automobile fire. Using a Class BC extinguisher, aim the nozzle at the engine fire and smother the flames with a series of short blasts.

Fire prevention is easier than fire fighting. Keep your boat's engine in good repair, don't allow it to become covered with a layer of grease that could catch fire, and don't keep a bunch of greasy rags lying around. When you refill the gas tank, let the engine cool first and use a funnel so you won't spill gasoline all over the place when a sudden wave tips the boat.

No boat should ever be taken away from a dock without having at least one life preserver per passenger aboard. Children should wear life jackets. If your boat is a fairly large one and you're accustomed to sailing it at sea or far out into a large lake, you should have a supply of signal flares at hand. A fire extinguisher is a must. One may not be enough; adjust the number according to the size of the boat. And here is a case where it is absolutely essential that every member of your party knows what to do in an emergency.

Fires don't occur only on power boats, of course. Some

folks manage to create a fire hazard even where one doesn't "naturally" exist. I find it hard to believe, but it happens: People actually use open charcoal grills aboard boats! There they are, a half-mile from shore, with a lake wind blowing a cloud of live coals around them and the boat being rocked this way and that by waves. Add some alcoholic beverages and a few nonswimmers and you have an almost inevitable disaster. Just because it's vacation time, don't leave your common sense at home!

Public Transportation

There's little I can tell you about fire danger aboard buses, trains, and airplanes other than that fire does happen, and that the best thing you can do is stay calm and *listen* to the people in charge. Buses and some train cars have emergency exits; take a seat close to one sometime and read the instructions that tell how it opens. Bus drivers usually have a fire extinguisher handy and have been trained to use it.

Restaurants, Theaters, and Public Buildings

Fire prevention and fire protection in public places such as restaurants, theaters, museums, and office buildings are usually aided by local ordinances that specify the legal capacity of the facility, the number and type of exits, and the placement of extinguishers and fire hoses. That still doesn't prevent common sense and even legally dictated rules being broken. It isn't at all uncommon to find exits chained shut, doors hidden behind stacks of cartons, or fire extinguishers that have been used and never recharged. Despite all the publicity surrounding spectacular and catastrophic fires in recent years, many buildings still don't have sprinkler systems or even smoke and heat detectors.

Make it your practice to look around when you enter a restaurant or a theater. Locate the emergency exits—and check them to make sure they're clear. If you find an exit locked or blocked, leave.

If you work in an office building, do you *really* know where all the stairs are? (If you ride an elevator up to your floor every day, you may have no idea where alternate exits are.) Do you know whether your building has sprinklers or smoke alarms? Take a few minutes *today* and find out.

AFTERWORD

I've pointed out some problems about home fires and have made some suggestions for you to act upon. I hope I've scared you a little. Fire is very scary. It scares me, and I've been around it for thirty years as a professional fire fighter.

I hope I've also reassured you a little and shown you that you don't have to be a fire victim. Most of all, I hope I've convinced you that you have to assume the responsibility for protecting yourself and your family against a home fire. *You* have to set up a plan of survival, and *you* have to see that it's learned, practiced, and perfected.

Let me tell you about an incident that happened recently in my own apartment building. I mentioned earlier that I live in a high-rise building. Fortunately, it's a building whose management believes in preparedness.

Because of my background, I was asked to speak to my neighbors about fire prevention. As it turned out, a couple of weeks after I finished a series of three evening discussions about fire prevention and survival, my instructions were put to the test.

One of my neighbors—I'll call her Mrs. Johnson—smelled something burning. Instead of panicking, she stopped for just a few seconds and thought about our tenants' meeting fire discussions. Then she decided upon her best course of action. The first thing she did was to call out to her husband, who was in another room. As she alerted him, she began feeling the entry door to her apartment. It seemed cool, so she cautiously opened it and looked out. The hallway was still clear, with just a few wisps of smoke beginning to gather near the ceiling. Since our building has self-closing, self-locking doors, Mrs. Johnson stationed her husband at their apartment's entry while she went across the hall to check the stairwell.

By this time, both the Johnsons could hear the sound of warning sirens. Mrs. Johnson could neither see nor smell any smoke in the stairwell, so she and her husband took their key and began the descent to street level—forty-one floors below!

The Johnsons were lucky—and smart. They arrived at the ground floor safely to learn that the fire had been a minor one: Someone had dropped a lighted cigarette down the rubbish chute. Smoke from the resulting small fire had climbed the shaft and had flowed out into some hallways through chute doors that had not been tightly closed. That was the lucky part: The fire was a minor one and was quickly extinguished without any extensive damage.

More important, though, the Johnsons knew what to do in a fire emergency *and they did it.* Every step of the way they kept calm and concentrated on getting quickly and safely away from danger.

Let's just take a look at the way they reacted: (1) Mrs. Johnson assessed the situation and took a few seconds to organize her thoughts; (2) she alerted her husband *while* she began moving to safety; (3) she checked the door for warmth before opening it; (4) she braced herself and *slowly* opened the door to check the corridor; (5) seeing only a little smoke, she decided the best thing to do was to leave the building; (6) by having her husband hold the door, she made sure she would not be trapped in the hallway; (7) instead of calling for the elevator, she proceeded to the nearest stairway and made sure it was clear of smoke; (8) Mr Johnson took the door key with him so that, if necessary they could return to the relative safety of their apartment (9) both the Johnsons began the long descent *by stairs* avoiding the elevator and possible entrapment; (10) by keeping their minds occupied with the business at hand—getting away from the danger—the Johnsons avoided panic.

The time from when Mrs. Johnson smelled smoke until they were both starting down the stairs was probably only a few seconds more than the time it took you to read the preceding paragraph. My point is, of course, that with the proper preparation and planning, you can train yourself *and your family* to react automatically to fire danger in your home. It's not difficult. It's not expensive. It doesn't take a lot of time

There's nothing mysterious or fancy or complicated about it. All you have to do is *right now* work out your family's fastest and simplest steps to safety. There's nothing I'd like better than knowing that you and your family are *prepared to stay alive!* Wouldn't you like that too?

A little five-year-old girl died one day because she was afraid to drop from a second-floor window onto a porch roof where we could get to her. No one had taken the time to show her what to do in a fire. No one had practiced getting out of her house with her. No one had installed a smoke detector that would have warned her of danger earlier so that she could possibly have found another way out.

That little girl's family probably still thinks of her often. They probably wonder what she would have been like if she had been allowed to grow up. And they probably still feel guilty about not having done something before it was too late.

I'd hate to live with that feeling.

SUGGESTED READINGS

Check under "fire" in the *Readers' Guide to Periodical Literature* at your local library for recent magazine articles on fire prevention, smoke detectors, fire extinguishers, and planning home fire drills. Examples of some very helpful articles are "Fire Extinguishers" (*Consumer Reports*, October 1979); "Beware the Wrong Home Fire Extinguisher" (*Changing Times*, December 1980); "Smoke Detectors" (*Consumer Reports*, August 1980); "Escape Ladder Improved" (*Consumer Reports*, February 1982); "Escape Ladders" (*Consumer Reports*, October 1979); "What to Do in a Hotel Fire" (*Good Housekeeping*, May 1981).

Some private companies and government agencies are good sources of information. Here are some to begin with.

A B C D Spells Fire. Free. Enclose a self-addressed stamped envelope. A clearly written pamphlet that explains in text and photographs how various types of fire extinguishers operate. Write to:

Underwriters Laboratories
Publications Division
333 Pfingsten Road
Northbrook, Illinois 60062

People & Fire. Free. Latest ideas on fire safety for homes, apartments, and mobile homes. An excellent all-around guide for planning a home fire prevention/fire evacuation program for your family. Write to:

*U.S. Department of Housing and Urban Development
Office of Policy Development and Research
Division of Energy, Building Technology, and Standards
Washington, DC 20410

* Note: There is a charge for some materials ordered from the Government Printing Office. When ordered directly from the publishing agency listed, however, each of these pamphlets and booklets is free.

Fire in Your Life (A Catalog of Flammable Products and Ignition Sources). Free. An excellent backup manual to use in educating your family to be fire-conscious. Tells how and why common household appliances and equipment can become hazardous. Write to:

*U.S. Consumer Product Safety Commission
Directorate for Communications
Washington, DC 20207

What You Should Know about Smoke Detectors. Free. Tells how various types of smoke detectors work and where they should be placed. Well illustrated. Write to:

*U.S. Consumer Product Safety Commission
Directorate for Communications
Washington, DC 20207

Home Fire Protection: Fire Sprinkler Systems. Free. Describes how home sprinkler systems work, method of installation, and cost. Write to:

*U.S. Fire Administration
Federal Emergency Management Agency
Washington, DC 20472

Fire in the United States. Free. A thorough and easy-to-read statistical breakdown of where (type of residence, which rooms, area of country), when (time of day, season), and why (heating, cooking, smoking, arson) most fires occur. Write to:

*U.S. Fire Administration
Federal Emergency Management Agency
Washington, DC 20472

Heating with Wood. Free. Offers tips for safe and efficient operation of fireplaces, stoves, and furnaces. Describes floor protection, stove placement, chimney cleaning, and the best types of wood to use. Write to:

*U.S. Department of Energy
Washington, DC 20585

How to Deal with Motor Vehicle Emergencies. Free. Describes actions to take in the event of an automobile fire—plus how to deal with a variety of other problems. Sized to fit into a glove compartment. Write to:

*U.S. Department of Transportation
National Highway Traffic Safety Administration
Office of Public Affairs and Consumer Services
Washington, DC 20590

After the Fire: Returning to Normal. Free. Tells how to assess damage, what legal steps to take, how to replace personal records. Tips on cleaning and salvaging belongings. Write to:

*U.S. Fire Administration
Federal Emergency Management Agency
Washington, DC 20472

You're Big Enough for Fire Safety (An Educational Unit for the Elementary Grades). Free. A total package that includes a teacher's guide, bulletin-board material, duplicating masters, games, and puzzles. Could be used by early-grade schoolteachers or community-center group leaders. Inquire at local Burger King restaurants or write to:

Burger King Fire Safety Program
G.P.O. Box 1472
New York, New York 10001

This is only a sampling of material available to the public. Your local fire department, insurance company, and gas or electric utility company may have more. Call them! Fire prevention is one area in which you can't become too educated. Read everything you can get your hands on—and then take action!

Often a booklet or information that could save your life is no farther away than a one-minute telephone call. Just to start you off: If you have a question about ignition sources or flammable products, call the U.S. Consumer Product Safety Commission in Washington. The number is (800) 638-2666 ([800]492-2937 if you live in Maryland), and the call won't cost you a cent.

INDEX